An Altitudinal Study of the Flora of the Inland Mountains of South-East Greenland

MONOGRAPHS ON GREENLAND
Vol. 357

BIOSCIENCE
Vol. 61

Geoffrey Halliday

An Altitudinal Study of the Flora of the Inland Mountains of South-East Greenland

Describing the Ammassalik – Kialiip Tasiilaa and Kangersertuaq Areas between 66° and 69° N

MUSEUM TUSCULANUM PRESS

An Altitudinal Study of the Flora of the Inland Mountains of South-East Greenland
Describing the Ammassalik–Kialiip Tasiilaa and Kangersertuaq Areas between 66° and 69° N
Geoffrey Halliday

© 2019 Museum Tusculanum Press and the author
Layout: Erling Lynder
Typesetting, composition and cover design: Janus Bahs Jacquet
ISBN 978 87 635 4554 9

Monographs on Greenland | Meddelelser om Grønland, vol. 357
ISSN 0025 6676

Bioscience, vol. 61
ISSN 0106 1054
Series editor: Reinhardt Møbjerg Kristensen
www.mtp.dk/MoG

This volume was originally submitted for publication in 2004.

Cover photo: Close-up of the grass *Poa abbreviata*, previously unknown south
of Scoresby Sund, taken at Bessel Fjord at 77° N in north-east Greenland.
Photo: Geoffrey Halliday, 1980

Museum Tusculanum Press
Dantes Plads 1
DK – 1556 Copenhagen V
Denmark

www.mtp.dk

Contents

Dedicated to the memory of Lt Cdr C.M. Stocken D.S.C., R.N.
killed while botanising in Schweizerland, 24 August 1966

Plate 1. The Northern area at Kangersertuaq (Kangerlussuaq): View north across a tributary cirque glacier on the east side of the Rosenborg Gletsjer (left). The cliffs of horizontally-bedded basalt, probably a fault scarp (Mathews 1979) rise almost 2000 m above the glacier. Locality A5 is the exposed rocks and the base of the adjacent cliff at the north-west end of the glacier. (Geodætisk Institut)

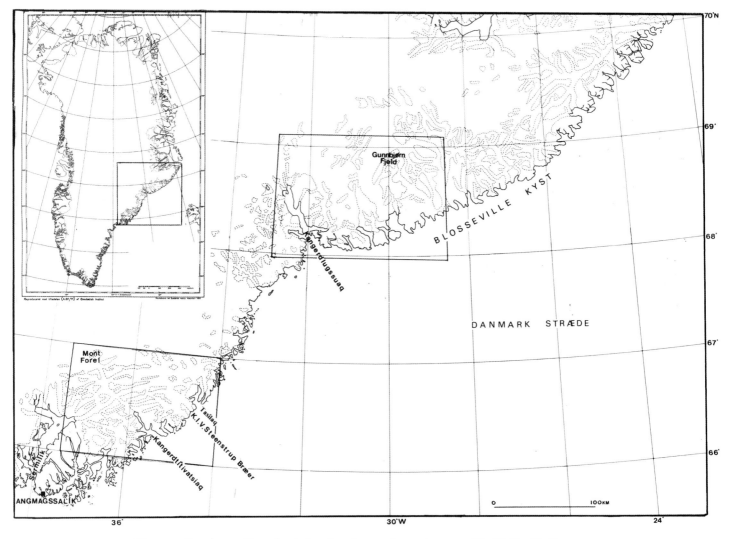

Figure 1. South-east Greenland showing the southern, Ammassalik–Kialiip Tasiilaa (Tasîlaq) and northern, Kangersertuaq (Kangerlussuaq) study areas featured in Figure 3 (p. 26) and Figure 4 (p. 34).

Introduction

This paper aims to bring together information from a variety of sources on the vascular flora of the botanically little-known-mountains and nunataks of south-east Greenland between Ammassalik and the Blosseville Kyst (66°–69° N) (Figure 1).

The paper is based on an analysis of a collection of mainly vascular plants from 111 individual sites of known altitude in the inland mountains between Ammassalik and Kialiip Tasiilaa (Tasîlaq) in the south, and around Kangersertuaq (Kangerlussuaq) in the north. The collections were made by 17 predominantly climbing expeditions, with little or no botanical experience, and all but one from 1963 to 1992. None of the sites had previously been visited, and the data are practically all unpublished except as botanical reports in general expedition reports which, for a variety of reasons, are now difficult or almost impossible to consult.

Although the climate on the coast can be characterised as low-arctic it becomes progressively drier and colder (i.e. high-arctic) inland. A consequence of this is the presence in the interior of the following species, none previously reported in east Geenland from south of Scoresby Sund (72° N): *Minuartia stricta*, *Melandrium affine*, *Draba arctica*, *D. subcapitata*, *Festuca baffinensis* and *Poa abbreviata*. There is also a cluster of records of *Arabis arenicola*, otherwise extremely rare in south-east Greenland.

100 vascular plants were recorded from the southern area and 86 from the northern. The highest sites were 2480 m in the south and 2100 m in the north, the former being the highest recorded in the Arctic. These altitudes were somewhat lower than those calculated by regression analyses of the number of species plotted against altitude. It seems very likely that many of the species occurring over about 1500 m reach their highest recorded altitudes in the Arctic within the study area.

The species are tabulated in order of increasing altitude limits and the altitudinal range is arbitrarily divided into three zones: low 370–1200 m, middle 1200–1800 m and high, above 1800 m. Low-arctic oceanic montane and low- and medium-arctic species preponderate in the lowest zone but with increasing altitude the widespread arctic montane element becomes dominant.

Comparison with other areas of Greenland shows that the altitudinal limits are noticeably higher in south-east Greenland.

The contrasting geology has no detectable affect on the flora, the best developed vegetation being on south-facing sites with summer melt water.

Less exhaustive lists of bryophytes and lichens are included in the Appendices.

I am greatly indebted to the many leaders of the expeditions featured in this paper for their willingness to co-operate with the botanical collecting and also to those who actually spared time to engage in what was for many of them a novel experience. Thanks are also due to Dr C.K. Brooks, Dr B. Fredskild, Dr E. Hoch and Dr E. Steen-Hansen for comments and information and to Ruth Berry for technical assistance with the figures.

Place names primarily used in this account are those authorised by the Greenland Language Secretariat. Many differ to a greater or lesser degree from the older names used in the maps, which were provided by Geodætisk Institut. Where relevant, these older names are given in parentheses after the new ones the first time they appear in a chapter, with one exception. The current name of Ammassalik is Tasiilaq, which name is used in the maps to refer to what is now called Kialiip Tasiilaa. In order to avoid confusion between these two locations, I use the older and more familiar name Ammassalik throughout.

Plate 2. Lemon Bjerge (northern area), comprising Domkirkebjerget (left) and the twin peaks of Mittivakkat (Mítivagkat, left), and the upper Frederiksborg Gletsjer viewed from the south. Localities C2 and C3 are on the north side of the former, B1 is at the south-east corner of Mittivakkat close to the glacier and B2 is on the western peak at 2100 m. W40, Gaffelen, is on the extreme right of the glacier. (H.G. Wager)

Chapter 1
Topography and geology

The whole area between Ammassalik and the Blosseville Kyst is extremely mountainous. In the Ammassalik area there are numerous rugged islands and an intricate system of fjords which penetrate up to 90 km from the outer coast. From south to north the main fjords are Sermilik, Kangersertivattiaq (Kangertítivatsiaq) and Kangersertuaq (Kangerlussuaq), all of which have major glacier systems calving into them: Nigertiip Apusiia ('Midgårdgletsjer'),

Plate 3. The southern area behind Ammassalik looking north-west. 16 September Gletsjer is in the foreground, from lower right to middle left. Nigertiip Apusiia ('Midgårdgletsjer') is in the middle distance from right to left. Beyond it on the left is Schweizerland and on the right is Femstjernen with its radiating glaciers. Mont Forel (3360 m) is in the far distance on the right. The collecting localities are spread over most of the area and inland almost to Mont Forel. (Geodætisk Institut)

Kattilersorpia ('Glacier de France') and Kangersertuaq Gletsjer respectively. The largest glaciers actually calve on the outer coast, for example, the parallel glaciers of K.I.V. Steenstrup Nordre and Søndre Bræer, and Christian IV Gletsjer, the snout of which is 15 km wide.

The coastal mountains mostly rise to between 700 and 1200 m. Inland the mountains increase steadily in altitude. Many exceed 2000 m and the highest lie close to the inland ice, these include Mont Forel (3260 m) in the south, and Gunnbjørn Fjeld (3693 m) in the north, the highest mountain in Greenland.

The geology of the area, based on studies by L.R. Wager (1947), C.K. Brooks (in litt.) and colleagues, can be briefly summarised as follows. Gneissic rocks of the Proterozoic Nagssugtoqidian orogenic complex dominate the area between Sermilik in the south and Sorgenfri Gletsjer east of Kangersertuaq. Tertiary intrusions related to the opening of the North Atlantic Ocean outcrop in its northern part, comprising gabbros and gabbro-like rocks, such as the Skærgård Intrusion on the east side of Kangersertuaq, and syenites and related rocks, on the west side of that fjord and in the Lilloise Bjerge. North-eastwards, from Sorgenfri Gletsjer to Scoresby Sund, is the large Tertiary basalt province which includes Watkins Bjerge. The original maximum thickness of the basalt is estimated to have been 6–7 km, thinning westwards, where the nunataks of Prinsen af Wales Bjerge, inland from the head of Kangersertuaq, are remnants of the plateau basalts. Cretaceous-Paleocene sediments, deposited in the Kangersertuaq Basin on the gneiss, and subsequently covered by the

Plate 4. View north-west towards the snout of the Kattilersorpia ('Glacier de France') (southern area). (D. Fordham)

Plate 5. Sortekap range (northern area) from the south, on the west side of Sorgenfri Gletsjer. Locality W3 is the low-lying exposure of sedimentary rocks on the right. (H.G. Wager)

basalt around the time of the continental break-up (55 Ma), are exposed on both sides of Sorgenfri Gletsjer and Christian IV Gletsjer. A few marine invertebrates, and some terrestrial plant fossils and coal, were reported from them (Courtauld 1936, Wager, L.R. 1937, 1947, Seward & Edwards 1941). More recently large numbers of late Cretaceous echinoids and molluscs, and remains of the mollusc-eating shark *Ptychodus decurrens* (Hoch 1991, 1992a), have been collected as well as late Cretaceous-Paleocene plant parts, such as leaves, cones and infructescences, of *Macclintockia, Ginkgo, Nordenskiöldia* and *Paranymphaea* as well as the well-known *Metasequoia* and *Cercidiphyllum* (Hoch 1991, 1992b). The youngest sedimentary rocks are considered to be remnants of the North Atlantic Land-bridge, which formed as part of the rifting process.

The resulting scenery is of three main rock types. The readily weathered, horizontally-layered basalts produce steep-sided dark mountains. These are flat-topped north of the Kronborg Gletscher (Plate 1), but more generally pyramidal or knife-edged (arêtes), due to faulting and glacial erosion, between the Kronborg Gletsjer and the Sorgenfri Gletsjer (Plate 1). The gneiss shows smoothly rounded or typical alpine topography (Plate 2, Plate 3, Plate 4) with steep, polished sides rising abruptly above the glaciers and often culminating in impressive arêtes and aiguilles. The sedimentary rocks (Plate 5, Plate 6) form low ridges and mountains supported by skeletons of basaltic dykes and shielded by harder rocks against glacial and other denudation.

The flora appears to be unaffected by these contrasting rock-types and their chemical composition. Plants occur wherever the physical conditions are favourable; they can be found on fairly fine, stable scree, on moraines, in gullies, on rock ledges and fissures and even on summit fell-field. The best-developed vegetation is usually south- facing and a really vital factor is an adequate supply of moisture from melting snow and ice throughout the growing season.

Plate 6. View north-westwards along the 'Sedimentary Ridge' with Sortekap in the distance (right). (E. Hoch)

Chapter 2
Climate

The main features of the climate of south-east Greenland have been discussed by Böcher (1938) and by Manley (1938), who analysed at some length the meteorological data of the 1935–6 British East Greenland Expedition based in the outer part of Kangersertuaq (Kangerlussuaq). There is, however, little factual information available on the climate of the inner fjords and mountains and one has to rely on personal recollections and brief comments in expedition reports.

The coast between Ammassalik and the Blosseville Kyst has a low-arctic climate. The February and July means for Ammassalik are -6°C and 7°C respectively. It is not uncommon in the winter for the temperature to rise briefly above zero, probably as a result of föhn winds. During the period 1954–1958, for example, 10% of the January readings were positive. The annual precipitation at Ammassalik is about 800 mm and both here and at Kangersertuaq precipitation is heaviest in October and November with a secondary peak in March and April. Of the Greenland weather stations only Prins Christian Sund (60° N) has more frequent precipitation than Ammassalik. Fog is frequent on the outer coast, especially in the north, but seldom penetrates far into the fjords. It can cause very marked temperature inversions. A recent geological expedition to the Blosseville Kyst found near-freezing temperatures in the fog by the shore and 'bikini weather', literally, on the sunlit ridges 1000 m above. Cold winds off the ice can dramatically lower summer temperatures and the cloudiness of the outer coast means that the snow-line is low. Böcher estimated this to be only 250–500 m in the outer parts of Kangerlussuaq and he considered this to be about the lowest in Greenland. However, this estimate is probably too low. Recent figures for Ammassalik, the Steenstrup Bræer and Kangersertuaq range from 600 to 900 m but perennating snow-drifts are common down to sea-level. The summer weather is notoriously changeable and varies greatly from year to year, being particularly cold and unsettled in bad ice years.

Most visitors to the interior agree that the summers are warmer, sunnier and drier than on the coast, although the weather is still rather unsettled and certainly not as good as in the interior of the fjords of central east Greenland. Although the interior mountains are higher than those on the coast, this is to some extent compensated for by a more favourable microclimate. While summer shade temperatures in the mountains may be near or below freezing, the high radiation in the dry air from surfaces exposed to the sun can result in temperatures of 20°C near the ground and presumably higher temperatures may occur on the generally south-facing screes, gullies, rock-ledges and fissures where most of the plant life occurs. It is interesting that the 1978 Westminster Expedition Report commented on the location of ivory gull nests on the cooler north-facing cliffs. Even in summer there may be periods of heavy snowfall: the 1966 Royal Naval Expedition Report, for example, recorded 1.4 m of snow in three days, but the amount accumulating on vegetated sites is likely to be small and to persist for only a short while. The interior is usually windier than the coast and several expeditions have commented on the gale-force winds and spin-drift which have seriously impeded sledging journeys. These gales often last for several days and may occur when the sky is clear; in Manley's (1938) opinion they are usually katabatic, only rarely being true föhn winds.

Such a transition from a maritime climate on the outer coast to a continental one in the inner fjords is characteristic of most parts of Greenland whether they lie within the low-arctic or high-arctic zones. Figure 2 is an attempt to represent this gradient diagrammatically.

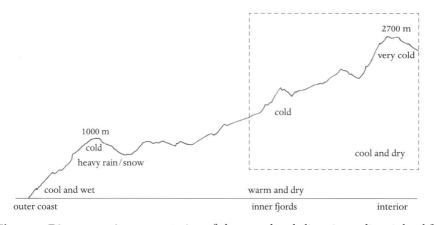

Figure 2. Diagrammatic representation of the postulated climatic gradient inland from the coast in the southern area, the area being indicated by the dashed lines.

Chapter 3
History of botanical exploration

Most botanical expeditions to Greenland have been concerned with the usually narrow strip of ice-free lowlands along the outer coast and the fjords. Collections from the mountains have been few and rarely from more than a few kilometres inland. This is certainly true of south-east Greenland as is amply demonstrated by the maps of botanical collecting sites between Kap Farvel and Scoresby Sund, and of the Ammassalik and Kangersertuaq (Kangerlussuaq) areas published by Böcher (1938, Figs. 1, 13 and 15). This neglect can be attributed partly to a lack of enthusiasm for mountaineering on the part of the botanists but it is largely due to a natural desire to explore as much of the coastal lowlands as possible in what little time ice conditions allow.

A notable exception to this general picture was the British East Greenland Expedition of 1935–36 (Wager 1937), an expedition which combined both scientific and mountaineering expertise and interests. It was based on the Skærgårdshalvø peninsula on the east side of Kangersertuaq (Figure 4) and it had as its chief purpose the detailed geological investigation of the Skærgård nephelene syenite igneous intrusion, discovered by the leader of the expedition, L.R. Wager, during the 1932 Scoresby Sound Committee's 2nd East Greenland Expedition led by Einar Mikkelsen. Wager's expedition included his brother, the plant physiologist H.G. Wager, and their wives. They overwintered on the Skærgårdshalvø and the four of them made very comprehensive and important botanical collections and observations over a wide area extending up the coast to Nansen Fjord and inland to the northern end of the Lindbergh Fjelde, 110 km from the outer coast. They also made the first ascent of Gunnbjørn Fjeld, 3693 m. A major journey on which they visited numerous nunataks in the interior was made each summer and it is the botanical results of these journeys, and of another to mountains on the west side of Kangersertuaq, which are included in this paper. The only other expedition to have collected in this area on a comparable scale was the Watkins Mountains Expedition of 1971. This expedition travelled up the coast from Ammassalik and, 36 years after the Wagers', climbed its objective, Gunnbjørn Fjeld, via the Rosenborg Gletsjer. Later expeditions benefitted from being landed directly onto the icecap: in 1988 the British East Greenland Expedition collected from a site just north of the Watkins Bjerge, the 1990 Northern Group Expedition landed in the northern part of the Kronprins Frederik Bjerge and made a small collection of lichens and moss from Panoramanunatakker, visited earlier by the Wager expedition, and two years later the British Mountaineering Expedition made a number of small collections from the general area of the Lemon Bjerge (Plate 2, p. 11), high up the Frederiksborg Gletsjer. Five montane collections were made by a 1932 Norwegian expedition, the 1972 Westminster East Greenland Expedition and a 1975 University of Copenhagen expedition.

The exploration of the mountains in the southern part of the area, north and north-east of Ammassalik (Figure 3, p. 26, and Plate 3, p. 12), did not really commence until 1963 when the area was visited by two climbing expeditions. One was from St Andrews University and led by the physicist P.W.F. Gribbon. He is also a keen botanist and on his own initiative he made extensive collections from 11 inland sites with the intention of studying altitudinal zonation. They were from an area christened by the expedition the 'Caledonian Alps' and lying between the 16 September Gletsjer and the heads of the Sermilik and Sermiligaaq fjords. He subsequently published his observations (Gribbon 1968a) and followed this up with a later paper comparing these findings with other data from nunataks in south and west Greenland (Gribbon 1968b). At the same time a Swiss-German expedition was active in much the same area making 15 collections of which half were from the 'Caledonian Alps' and half from north of the 16 September Gletscher.

Three years later, in 1966, two more climbing expeditions visited the area, a Royal Naval expedition and one from Switzerland. The latter was led by S. Angerer, a member of the earlier Swiss-German expedition, and collections were made over a wide area but chiefly on either side of the Kattilersorpia ('Glacier de France') (Plate 4, p. 13) which comes down to the coast at the head of Kangersertivattiaq (Kangertítivatsiaq). The ill-fated Naval expedition was led by C.M. Stocken (Plate 8, p. 28), a keen amateur botanist. Collections were made from 11 sites lying north of the 16 September Gletsjer and extending north-west into Schweizerland. It was while descending a mountain in Schweizerland towards the end of August that Stocken fell and was fatally injured.

In 1967 an Imperial College expedition ascended Kattilersorpia and made four collections from mountains far inland between Mont Forel and Femstjernen. The following year an Army expedition explored the mountains around Kristian Gletsjer, east of Femstjernen, and collected from two more sites. Finally, in 1969, a London University Graduate expedition climbed around the same glacier and made three collections from its southern side.

Attention then moved northwards and in 1971 an expedition to the Steenstrup area travelled up the coast and after visiting Kangersertivattiaq continued north to Ikertivaq establishing a base camp at the head of Kialiip Tasiilaa (Tasîlaq), its main branch, and making a collection from a nunatak 11 km to the north-west. Three more expeditions were to visit the same area, the 1974 and 1978 Westminster expeditions (Woolley 2004) and the 1980 Durham University expedition. All camped near the head of the fjord and collected on journeys into the interior. From Kialiip Tasiilaa the first Westminster expedition travelled 60 km to the north-west and the Durham expedition 35 km in the same general direction, but the second Westminster expedition made an epic journey of 140 km northwards through the inner nunataks of the Kronprins Frederik Bjerge to 68° N. The scientific work of this last expedition was largely confined to important observations on the ivory gull (*Pagophila eburnea*) and the collection of lichens. A few specimens of vascular plants were collected but unfortunately the material was either scrapped or lost and no identifications are available.

Information on all the above expeditions is summarised in Table 1. The plant collections of only four of these were made by professional botanists. Those of a further two were made by keen amateurs, but the majority were made by individuals with no botanical training and who agreed, sometimes reluctantly, to burden themselves with a plant press. They were asked to collect samples of all the vascular plants at a site. Obviously the collections are likely to be less comprehensive than one would like and monocotyledons in particular are certainly under-represented. Yet the large number of collections available makes it likely that very few species have been completely overlooked. It is important to realise that the data presented in this paper do not represent the results of a rigorous scientific study of altitudinal ranges resulting from sampling at specific altitudes. Collections were made as and when expedition members

YEAR	NAME	LEADER
1933	(Norwegian Expedition)	P.S. Brandal
1935–36	British East Greenland Expedition	L.R. Wager
1963	Scottish East Greenland Expedition	P.W.F. Gribbon
1963	Schweizerische-Deutschen Grönland Expedition	S. Angerer
1966	Royal Naval East Greenland Expedition	C.M. Stocken & R.H. Wallis
1966	Schweizer Grönland-Expedition	S. Angerer
1967	Imperial College East Greenland Expedition	R. Chadwick
1968	Army East Greenland Expedition	C.H. Agnew
1969	London University Graduate Mountaineering Club East Greenland Expedition	D. Fordham
1971	Expedition to the Steenstrups area of East Greenland	M. Tuson & H.J.D. Webster
1971	Watkins Mountains Expedition Phase II	A.J. Allan
1972	Westminster East Greenland Expedition	E.R.D. French
1974	Westminster East Greenland Expedition	W.S.L. Woolley
1975	University of Copenhagen	C.K. Brooks
1978	Westminster East Greenland Expedition	W.S.L. Woolley
1979	Durham University Polar East Greenland Expedition	W. Rigden
1988	British East Greenland Expedition	J. Lowther
1990	Northern Group Greenland Expedition	W.S.L. Woolley
1992	British Mountaineering Expedition to East Greenland	P. Bartlett

Table 1. Chronological list of expeditions.

*   Copies of all expedition reports and botanical MSS are in the possession of the present author

†   See also Woolley (2004)

‡   International Herbarium Abbreviation

| MAIN BOTANICAL COLLECTOR | PUBLICATIONS[*] | | HERBARIUM MATERIAL[‡] |
	GENERAL	BOTANICAL	
P.F. Scholander		Devold & Scholander (1933)	O
H.G. & L.R. Wager and wives	Courtauld (1936), Wager, L.R. (1937)	Scattered references in Böcher (1938), Growth and age studies in Wager, H.G. (1938); Wager, H.G. & Mrs L.R., & Wilmott, A.J. MSS	BM, E
P.W.F. Gribbon	Report	Report, Gribbon (1968a,b)	E, STA
S. Angerer et al.	Report	K. Holmen MS	C
C.M. Stocken	Report	G. Halliday in Report	C, E
S. Angerer et al.	Report	Lowland data in Elsley & Halliday (1971); G. Halliday MS	C
R.G. Swainson	Report	Lowland data in Elsley & Halliday (1971); G. Halliday MS	C, E
B.K. Porter	Report	G. Halliday in Report	C, E
D. Fordham	Report	G. Halliday in Report	BM, C, E
R. Coulter	Report	G. Halliday in Report	BM, E
A.J. Allan	Preliminary Report	A.J. Allan MS	BM
M.C. Burns et al.	Report[†]	Report	C, E
E.L. Lodge	Report[†]	Report	untraced
C.K. Brooks		G. Halliday MS	
R.E. Illingworth	Report[†]	Report (lichens only but some vascular plants were collected)	BM (lichens)
B. Hines & M. Wilson	Report	Report	C, E
G. Englefield	Report	Report	E
J.H. Richardson	Report[†]	G. Halliday in Report, lichens and mosses only	C, E
R.E. Illingworth	Report		

had the opportunity. No significance should be attached to the difference in altitudes of the lowest sites in the northern and southern areas.

This paper includes all the botanical records of these expeditions from montane sites down to an arbitrary limit of 370 m. Collections were of course made at lower altitudes, chiefly around the base camps by the inner fjords. Consideration of these is outside the scope of the present paper but, since it is important that the existence of these lowland records be more generally known, further information is given in Appendix I.

The vascular plant nomenclature used in this paper largely follows that of *Grønlands Flora* (Böcher et al. 1978).

Chapter 4
List of collecting localities

The localities are listed below under the relevant expedition and the major collector, the expeditions being arranged chronologically. Each locality is prefixed by a letter appropriate to the expedition and the original numbering of the localities has in nearly every case been retained. To facilitate reference to the expedition reports the original number, where different, is given in parentheses. Where localities were originally given letters, numbers have been substituted and the letters given in parentheses. A few localities visited more than once and at more or less the same altitude have been united. Omitted from the list are lowland sites and also a few montane ones with less than five species, except where such sites are clearly approaching the altitudinal limit for vascular plants. Also omitted are the five new sites visited by the 1978 Westminster expedition from which vascular plants are known to have been present; these are given in Appendix III. Unofficial place-names are in parentheses. The sites are shown on the maps of the southern and northern areas (Figure 3, p. 26, and Figure 4, p. 34).

Figure 3. The southern area showing the collecting localities.

4.1 The southern area: Ammassalik–Kialiip Tasiilaa (Tasîlaq) (Figure 3 and Plate 3, p. 12)

4.1.1 Scottish East Greenland Expedition 1963: P.W.F. Gribbon

G2   South side of mountain at head of Kaarali Gletscher, 66° 12′ N, 36° 50′ W, 940–1160 m, south-facing (see also D13); M. Lenarčič collected only a few species from this site in 1978, but three are additional to Gribbon's list: *Cassiope tetragona*, *Arnica angustifolia* and *Tofieldia pusilla*.

G3   Mountain north-east of head of Kaarali Gletsjer, 66° 12′ N, 36° 45′ W, nunatak at 930 m (locality 12) and south-facing slopes 960–1000 m.

G4   Mountain north-east of head of Kaarali Gletsjer, 66° 12′ N, 36° 47′ W, south~facing slopes, 1000–1150 m (see also G14).

G5   Slangen, between head of Kaarali Gletsjer and 16 September Gletsjer, 66° 14′ N, 36° 40′ W, south ridge, c. 1350 m.

G6   Mountain west of Knud Rasmussen Gletsjer, 66° 11′ N, 36° 20′ W, east-facing ledge, 1400 m.

G7   Rytterknægten, 66° 08′ N, 36° 42′ W, east-facing ledge, 1510 m (see also D12).

G8   Mountain west side of snout of Knud Rasmussen Gletsjer, 66° 05′ N, 36° 25′ W, south-facing ledges, 1560–1580 m.

G13   Rytterknægten, north side, 66° 09′ N, 36° 42′ W, west-facing, 1060 m, P.W.F. Gribbon.

G14   Mountain north-east of head of Kaarali Gletsjer, 66° 12′ N, 36° 47′ W, summits at 1385 and 1435 m (locality 15), P.W.F. Gribbon (see also G4).

G16   Puttuut (Trillingerne), south-east corner, 66° 07′ N, 36° 58′ W, south-facing, 810–830 m.

G17   Mountain south of 16 September Gletsjer, 66° 11′ N, 36° 33′ W, south-facing scree slopes and ridge, 1060–1160 m.

Plate 7. Schweizerland (southern area) viewed from Gardenbjerg (Point 2500 m) with Point 2480 m on the left and Stockenbjerg, 2520 m, in the centre. (T. Kirkpatrick)

Plate 8. Schweizerland (southern area). C.M. Stocken looking south-east from Point 2330 m across Franche Comté Gletsjer to Laupersbjerg; Femstjernen and Kattilersorpia ('Glacier de France') are on the extreme left. (T. Kirkpatrick)

4.1.2 Schweizerisch-Deutsche Grönland-Expedition 1963: S. Angerer et al. (Plate 3)

D1 Mountain north-east of head of Kaarali Gletsjer, 66° 12′ N, 36° 44′ W, 1370 m, 17/7/1963.

D2 (=S4) Mountain on south side of 16 September Gletsjer, 66° 15′ N, 36° 34′ W, 1050 m and 1100 m (locality 8), 19 & 30/7/1963.

D3 South-west corner of mountain on north side of 16 September Gletsjer, 66° 18′ N, 36° 38′ W, 900–950 m, 23/7/1963.

D4 South side of mountain to north of 16 September Gletsjer, 66° 20′ N, 36° 35′ W, 1500 m, 22/7/1963.

D5 East side of mountain to north of 16 September Gletsjer, 66° 19′ N, 36° 38′ W, 1900 m, 22/7/1963.

D6 As D5 but 2 km to south-west, 66° 18′ N, 36° 41′ W, 1680–1850 m, 23/7/1963.

D7 North-east corner of mountain on south side of 16 September Gletsjer, 66° 16′ N, 36° 46′ W, 1200–1650 m, 24/7/1963.

D9 North side of junction of 16 September Gletsjer and Knud Rasmussen Gletsjer, 66° 18′ N, 36° 18′ W, 1750–1800 m, 3/8/1963.

D10 South side of Håbet Gletsjer, 66° 20′, 36° 15′ W, 2000 m, 3/8/1963.

D11 Point 1400, east side of Knud Rasmussen Gletsjer, 66° 15′ N, 36° 08′ W, 1460 m, 4/8/1863.

D12 Rytterknægten, 66° 08′ N, 36° 42′ W, 1400 m, 15/8/1963 (see also G7).

D13 South side of mountain at head of Kaarali Gletsjer, 66° 12′ N, 36° 50′ W, 760 m, 16/8/1963 (see also G2).

D14 Point 1500, north-east side, 66° 05′ N, 36° 52′ W, 1050 m, 18/8/1963.

D15 Between Points 1306 and 1500, 66° 04′ N, 36° 57′ W, 1200 m, 18/8/1963.

D16 Point 1500 m, 66° 04′ N, 36° 53′ W, 1000 m, 19/8/1963.

4.1.3 Royal Naval East Greenland Expedition 1966: C.M. Stocken (Plate 7, Plate 8)

N4 Mountain at junction of Rasmussen Gletscher and 16 September Gletsjer. 66° 14′ N, 36° 17′ W, rocky cliffs, 700–800 m, 20/7/1966.

N5 Conniat Bjerg, 66° 23′ N, 36° 27′ W, east- to south-east-facing, 1600–1800 m, 23/7/1966.

N6   Mountain east of Sølverbjerg, 66° 22′ N, 36° 19′ W, south-facing cliffs, 1750 m, 25/7/1966.

N7   Mountain on west side of Femstjernen, 66° 38′ N, 36° 43′ W, south-facing cliffs, 1100 m, 29/7/1966.

N8   Mountain north of Devaux Bjerg, 66° 28′ N, 36° 19′ W, south-facing cliff ledges, 900–1700 m, 3/8/1966.

N9   Point 2000, 66° 25′ N, 36° 14′ W, rock ledges, 1700–2000 m, 9/8/1966.

N10   (= S3) Mountain on south-west side of Kattilersorpia, 66° 33′ N, 36° 25′ W, steep, south-east-facing cliffs, 800 m, 11/8/1966.

N11   Mountain on west side of Femstjernen, 66° 39′ N, 36° 47′ W, rocky col, 1300 m, 13/8/1966.

N12   Mountain south-east of Stocken Bjerg, 66° 36′ N, 37° 05′ W, south-facing slopes, 1800–2480 m, 16–18/8/1966. (Plate 7)

N13   Garden Bjerg, 66° 40′ N, 37° 10′ W, south face, 2200–2400 m, 21/8/1966. (Plate 8)

N14   Point 2200, 66° 38′ N, 36° 48′ W, north-facing rock ledges, 1900 m, 23/8/1966.

4.1.4 Schweizer Grönland-Expedition 1966

S1   Mountain on east side of Knud Rasmussen Gletsjer, 66° 09′ N, 36° 10′ W, north-west-facing, 370 m, M. Zumbuhl (B), 17/7/1966.

S2   Mountain east of Sermiligaaq, 66° 01′ N, 36° 08′ W, south-facing, 700 m, S. Angerer (C), 2/7/1966.

S3   (=N10) Mountain on south-west side of Kattilersorpia, 66° 33′ N, 36° 25′ W, south-east-facing, 880 m, S. Angerer (D), 7/8/1966.

S4   (= D2) Mountain on south side of 16 September Gletsjer, 66° 15′ N, 36° 34′ W, north-east-facing, 1020 m, M. Zumbuhl (F) 22/8/1966.

S5   Parat Bjerg, south end, 66° 26′ N, 36° 25′ W, south-east-facing block boulders, 1050 m, M. Zumbuhl (G), 23/7/1966.

S6   Point 1600, 66° 31′ N, 36° 02′ W, north-facing, 1100 m, H. Hagenbuch (H,J) 26/7/1966 (see also S12).

S7   Rødebjerg, south-west corner, 66° 27′ N, 36° 28′ W, south-facing scree, 1200 m, M. Zumbuhl (I), 18/8/1966.

S8   Mountain on north side of Pourquoi-pas Gletsjer, 66° 39′ N, 36° 00′ W, south-facing, 1350 m, H. Hagenbuch (K), 10/8/1966.

S9   Mountain on south side of Pourquoi-pas Gletsjer, 66° 39′ N, 35° 42′ W, south-facing, 1500 m, M. Zumbuhl (L), 8/8/1966.

S10   Mountain south of 16 September Gletsjer, 66° 12′ N, 36° 33′ W, south-facing, 1500 m, M. Zumbuhl (M), 22/8/1966.

S11   Mountain south of 16 September Gletsjer, 66° 13′ N, 36° 37′ W, south-facing, 1580 m, S. Angerer (N), 21/8/1966.

S12   Point 1600, 66° 31′ N, 36° 02′ W, north-west-facing, 1500 m, H. Hagenbuch (O), 26/7/1966 (see also S6).

S13   Point 2006, 66° 37′ N, 36° 30′ W, south-south-east-facing, 1780 m, H. Hagenbuch (P), 7/8/1966 (see also S15).

S14   Point 2100, 66° 35′ N, 36° 03′ W, south-east-facing, 2000 m, M. Zumbuhl (Q), 15/8/1966.

S15   Point 2006, 66° 37′ N, 36° 30′ W, south-facing 30°, 2000 m, H. Hagenbuch (R), 7/8/1966 (see also S13).

4.1.5   Imperial College East Greenland Expedition 1967: R.G. Swainson

I1   Mountain on east side of Paris Gletsjer, 66° 47′ N, 36° 44′ W, south-facing scree and ledges, 1470–1570 m, (A), 8/8/1967.

I2   Southernmost point of Avantgarden, 66° 49′ N, 36° 39′ W, damp, south-facing ledges, 1700–1800 m, (B), 10/8/1967.

I3   Mountain south of Avantgarden, 66° 49′ N, 36° 36′ W, west-facing ledges and gullies, 1800–1950 m, (C), 12/8/1967.

I4   Southernmost point of Henri-Dunant Bierg, 66° 51′ N, 36° 34′ W, south-facing ledges, 2050–2150 m, (D), 17/8/1967.

I5   North-east side of Kattilersorpia 14 km from the head of Kangerser-tivattiaq (Kangertítivatsiaq), 66° 31′ N, 36° 09′ W, damp, south-facing ledges and gullies, 600–650 m, (E), 22/8/1967.

4.1.6 Army East Greenland Expedition 1968: B.K. Porter

A1   (=L1) Mountain on south side of Kristian Gletsjer, 66° 46′ N, 36° 16′ W, north-north-east-facing rock crevices, 1670 m, (A), 19/6/1968.

A2   Southernmost point of Point 2730, 66° 48′ N, 36° 13′ W, south-facing loose earth and rubble, 1900 m, (B), 5/8/1968.

4.1.7 London University Graduate Mountaineering Club East Greenland Expedition 1969: D. Fordham

L1   (=A1) Mountain on south side of Kristian Gletsjer, 66° 46′ N, 36° 16′ W, loose, north-east-facing cliff, 1700 m, (A), 25/7/1969.

L2   Point 2000, 66° 43′ N, 36° 24′ W, south-facing scree, 2000 m, (B), 26/7/1969.

L3   Ridge east of Point 2000, 66° 44′ N, 36° 17′ W, south-facing scree, 2300 m, (C), 27/7/1969.

4.1.8 Expedition to the Steenstrup area of East Greenland 1971: R. Coulter

T1   (=W1) Breubjerg, 11 km north-west of Kialiip Tasiilaa, 66° 42′ N, 34° 40′ W, south- to south-east-facing rocky slopes, 1050–1200 m, (B), 15/8/1971.

4.1.9 Westminster East Greenland Expedition 1974: E.L. Lodge

W1   (=T1) Breubjerg, north-west corner, 11 km north-west of Kialiip Tasiilaa, 66° 42′ N, 34° 40′ W, north-facing loose cliff, 900–950 m, 24/7/1974.

W2   Nunatak east of Point 1800, 66° 42′ N, 34° 44′ W, south- to south-west-facing scree, 1050–1100 m, 25–27/7/1974 & 18/8/1974.

W3   Mountain on north side of upper K.I.V. Steenstrup Nordre Bræ, 66° 44′ N, 35° 04′ W, east-facing block scree, 1250 m, 1/8/1974.

W5   Mountain on east side of 'Seksstjernen', 66° 53' N, 35° 12' W, south-facing rocky ridge with broad ledges, 1500–1600 m, 2, 3 & 14/8/1974.

W7   Small nunatak north of 'Seksstjernen', 67° 01' N, 35° 50' W, west-facing scree, 2100–2150 m, 7/3/1974.

W9   Mountain on north side of 'Seksstjernen', 66° 54' N, 35° 34' W, steep south-west-facing sheltered gullies, 1550–1580 m, 10/3/1974.

W10   Mountain on north side of 'Seksstjernen', 66° 54' N, 35° 27' W, south-facing scree slope, 1490–1500 m, 11/8/1974.

W11   Mountain on north side of 'Seksstjernen', 66° 53' N, 35° 24' W, south-west-facing scree, 1450–1600 m, 11/8/1974.

W12   Mountain on east side of 'Seksstjernen', 66° 51' N, 35° 08' W, south- to south-west-facing steep sheltered gullies, 1500–1520 m, 15/8/1974.

W13   Mountain north-east of upper K.I.V. Steenstrup Nordre Bræ, 66° 45' N, 35° 02' W, south-east-facing scree, 1200–1400 m, 17/8/1974.

4.1.10 Durham University Polar East Greenland Expedition 1979: M. Wilson & B. Hines

M1   Point 2000, 66° 55' N, 34° 50' W, north-west-facing fell-field, 1600 m, 14/8/1979.

M2   Peak 2200, 66° 56' N, 34° 56' W, east-facing rocky slopes, 1500–2100 m, 15/8/1979.

M3   (= T1) Small nunatak 11 km north-west of Kialiip Tasiilaa, 66° 43' N, 34° 44' W, east- and north-east-facing fell-field, 900 m, 20/8/1979.

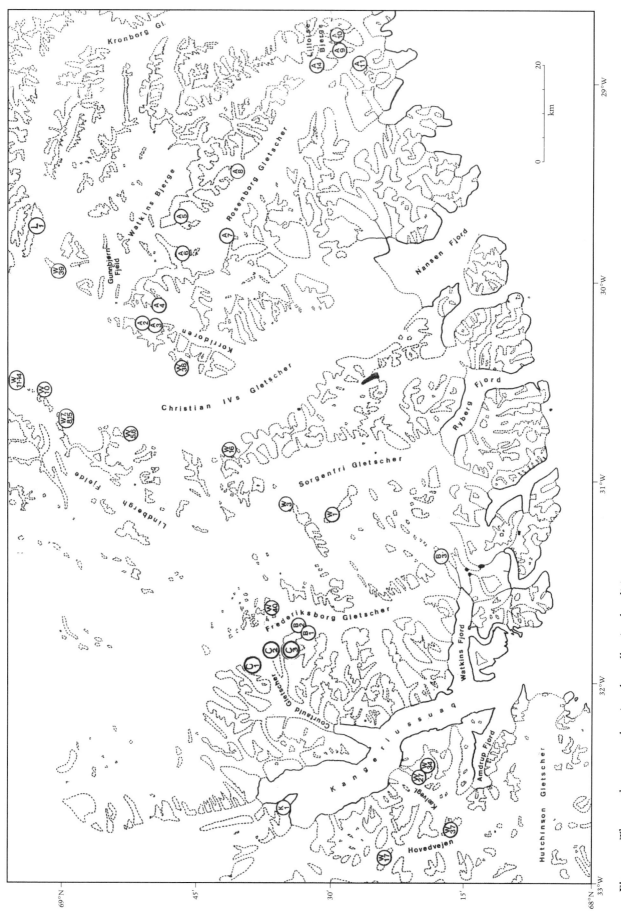

Figure 4. The northern area showing the collecting localities.

4.2 The northern area: Kangersertuaq (Kangerlussuaq) (Figure 4)

4.2.1 Norwegian Expedition 1932: P.F. Scholander

S1 'Brandalfjell', between Mudderbugt and Søndre Syenitgletscher, west side of Kangersertuaq, 68° 17′ N, 32° 18′ W, plateau and scree, 900–1060 m, 23/8/1932.

4.2.2 British East Greenland Expedition 1935–1936

W1 Ridge south-east of Sortekappasset towards Pyramiden, 68° 31′ N, 31° 10′ W, summit ridge and both slopes, 1300–1350 m, L.R. Wager et al. (SJ1,2,2a), 21/8/1935, 25/7/1936.

W3 North-east end of Sortekap range by Sorgenfri Gletsjer (Sedimentary ridge), 68° 35′ N, 31° 06′ W, 1300 m, H.G. & E.M. Wager (SJ3), 26/7/1936. Visited in 1988 by E. Hoch who added *Arabis arenicola* to the Wagers' list (Plate 5, Plate 6).

W5 Nunatak north-east of Kulhøje, Lindbergh Fjelde, west side of Christian IV Gletsjer, 68° 53′ W, 30° 45′ W, south-east corner, 1450 m, H.G. & E.M. Wager (SJ5), 30/7/1936.

W6 As W5 but gently sloping area at 1550 m (SJ6).

W7 East side of Lindbergh Fjelde, west side of Christian IV Gletsjer, 69° 01′ N, 30° 38′ W, low scree and south-facing moraine, 1400–1420 m, H.G. & E.M. Wager (SJ7,9), 31/7/1936.

W8 As W7 but at junction of screes and crags, 1600 m, (SJ8, 15a), 31/7 & 3/8/1936.

W10 Nunatak on west side of Christian IV Gletsjer, 2 km east of Lindbergh Fjelde, 69° 03′ N, 30° 30′ W, small isolated rock mass, 1440 m, H.G. & E.M. Wager (SJ10), 1/8/1936.

W11 North-east end of Lindbergh Fjelde, head of Christian IV Gletsjer, 69° 06′ N, 30° 27′ W, 1500 m, H.G. & E.M. Wager (SJll), 1/8/1936.

W12 As W11, south-east-facing rocks, 1600 m, (SJ12).

W13 As W11, south-west-facing rocks, 1830 m, (SJ13), 2/8/1936.

W14   As W11, south-west-facing rocks, 2000 m, (SJ14), 2/8/1936.

W15   As W7, 1760–1900 m, (SJ15b-d), 3/8/1936.

W16   Dumpen, west side of Christian IV Gletsjer, 68° 42′ N, 30° 50′ W, 1200–1400 m, L.R.Wager et al. (SJ16), 20/8/1935, 4/8/1936.

W17   Nunatak between Hovedvejsnunatakker and Point 1800 to north-east, west side of Kangersertuaq, 68° 24′ N, 32° 55′ W, 1400 m, P. Wager (SJ17), 10/7/1936. Included here is SJ18 collected on same day either from this nunatak or from Hovedvejsnunatakker 4 km to the south-west.

W27   Mountain on south-east side of 'Kælvegletscher' near junction with Nordre Syenitgletsjer, west side of Kangersertuaq, 68° 20′ N, 32° 32′ W, 760 m, P. Wager, 16/7/1936.

W34   Mountain on north side of Nordre Syenitgletsjer, 68° 19′ N, 32° 24′ W, 610 m, P. Wager, 2/7/1936.

W37   Trebjørnebjerget, east side of Hovedvejen glacier, 68° 16′ N, 32° 43′ W, 1200–1300 m, P. Wager, 9/7/1936.

W38   Mountain above Point 1210, Skærmen, east side of Christian IV Gletscher, 68° 47′ N, 30° 25′ W, 1250 m, H.G. & L.R. Wager, 19/8/1935.

W39   South end of nunatak, east side of Christian IV Gletsjer, north-west of Gunnbjørn Fjeld, 69° 01′ N, 30° 00′ W, c. 1500 m, H.G. & L.R. Wager, 15/8/1935.

W40   Gaffelen, head of Frederiksborg Gletsjer, 68° 37′ N, 31° 37′ W, 1300 m, E.C. Fountaine, 3/6/1936 (Plate 9).

4.2.3   Watkins Mountains Expedition
Phase II 1971: T. Eriksen et al.

A2   West side of Korridoren, 68° 51′ N, 30° 12′ W, east-facing, sheltered cracks in basalt cliff, 1890 m, (B), 8/8/1971.

A3   West side of Korridoren, 68° 50′ N, 30° 12′ W, morainic debris, 1950 m, (C), 8/8/1971.

A4   East side of Korridoren, 68° 49′ N, 30° 06′ W, west-facing, very weathered basalt moraine, 1740 m, (D), 9/8/1971.

A5   Mountain north-east of head of Rosenborg Gletsjer, 68° 47′ N, 29° 38′ W, south-facing, wet open slopes of weathered debris, 1460–1740 m, (E), 11/8/1971 (Plate 1, p. 6).

Plate 9. Gaffelen (northern area, locality W40) from the south. Frederiksborg Gletsjer is on the left and Lindbergh Fjelde is in the far distance on the right. (Geodætisk Institut)

A6   Low nunatak at head of Rosenborg Gletsjer, 68° 47′ N, 29° 51′ W, west-facing cracks in basalt, 1520–1710 m, (F), 12/8/1971.

A7   Mountain on south-west side of Rosenborg Gletsjer near its head, 68° 41′ N, 29° 45′ W, south-east-facing, cracks in basalt and on damp soil, 1370 m, (G), 13/8/1971.

A8   Mountain on north-east side of Rosenborg Gletsjer, 68° 40′ N, 29° 25′ W, south-facing, cracks in rock and on damp soil, 1190–1310 m, (H), 14/8/1971.

A9   South-west corner of Lilloise Bjerge, 68° 28′ N, 28° 48′ W, south-west-facing granite scree slope, 490–940 m, (I), 19/8/1971.

A10   South end of Lilloise Bjerge, 68° 29′ N, 28° 44′ W, granite ridge, 1280–1400 m, (J), 20/8/1971.

A11 Mountain on north side of glacier pass behind Kap Normann, 68° 26′ N, 28° 54′ W, south-facing rock ledges and cracks in basalt, 790 m, (K), 21/8/1971.

A14 Western end of Lilloise Bjerge, 68° 31′ N, 28° 54′ W, north-west-facing basalt scree slope, 850–1220 m, (N), 18/8/1971.

4.2.4 Westminster East Greenland Expedition 1972
(Plate 2, p. 10)

B1 South-east side of Mittivakkat, west side of Frederiksborg Gletsjer, 68° 33′ N, 31° 44′ W, 1300–1400 m, C.D.J. Kessler & P.J. Robinson (7), 15 & 17/8/1972.

B2 Near west summit of Mittivakkat, above head of Frederiksborg Gletsjer, 68° 34′ N, 31° 43′ W, damp, sheltered, south-facing ledge, 2100 m, C.D.J. Kessler & P.J. Robinson, 16/8/1972.

B3 Nunatak ridge 8 km east from head of Watkins Fjord, 68° 17′ N, 31° 24′ W, 1000 m, M.C. Burns (6), 16/8/1972.

4.2.5 University of Copenhagen Kangersertuaq Expedition 1975: C.K. Brooks

K1 Nunatak north-west of Batbjerg, head of Kangersertuaq, 68° 39′ N, 32° 49′ W, steep, south-facinq gully, 700–1000 m.

4.2.6 British East Greenland Expedition 1988: G. Englefield

L1 'Knud Rasmussen Land' north of Watkins Bjerge, south of Point 3018 m, 69° 04′ N, 29° 41′ W, damp, south-facing slope above the glacier, 2100 m, (1,3).

4.2.7 British Mountaineering Expedition to East Greenland 1992: R. Illingworth (Plate 2, p. 11)

C1   Mountain at head of Courtauld Gletsjer, 68° 39′ N, 31° 56′ W, south-east-facing summit boulders, c. 1950 m, (4), 24/7/1992.

C2   Northernmost mountain of Lemon Bjerge, 68° 36′ N, 31° 50′ W, steep east-facing boulder slope, c. 1500 m, (5), 26/7/1992.

C3   Domkirkebjerget (Point 2600), Lemon Bjerge, 68° 35′ N, 31° 50′ W, east-facing boulder scree, c. 1750 m, (7), 30/7/1992.

C4   (=W40) Gaffelen, upper Frederiksborg Gletsjer, 68° 37′ N, 31° 40′ W, south-west-facing boulders below basalt dyke, 1300 m, (6), 27/7/1992 (Plate 9).

Chapter 5
Analysis of the plant records

The plant records from all the sites listed in the previous chapter are presented in Table 2 (p. 43) and Table 7 (p. 61), the former covering the southern area (Figure 3, p. 26) and the latter the northern (Figure 4, p. 34). The sites are presented in order of increasing altitude and the species are arranged in the order of their altitudinal limits. The data for the two areas will first be considered separately and analysed under the following three headings:

1   The altitudinal ranges and distribution in south-east Greenland of the individual species.
2   The rate of decrease of the flora with altitude.
3   The effect of altitude on the floristic elements represented in the flora.

5.1   The southern area: Ammassalik – Kialiip Tasiilaa (Tasîlaq) (Figure 3, p. 26; Table 2)

5.1.1   The altitudinal ranges and distribution of species

The number of species in Table 2 – 100 – is remarkable amounting as it does to about 45% of those recorded from the whole of the Ammassalik area. The number per site varies greatly even allowing for altitude. This reflects the considerable variation in such habitat factors as aspect, exposure and water supply and to a lesser extent the varying botanical expertise of the collectors. The contrast between sites at the same altitude is well shown by Gribbon's data for sites G3 (960–1000 m) and G13 (1060 m). The former had the largest number

UPPER ZONE
- L3 2300m
- N13 2200–2400m
- W7 2100–2150m
- I4 2050–2150m
- L2 2000m
- S15 2000m
- D10 2000m
- S14 2000m
- D5 1900m
- N14 1900m
- A2 1900m
- N12 1800–2480m
- I3 1800–1950m

MIDDLE ZONE
- S13 1780m
- D9 1750m
- N6 1750m
- N9 1700–2000m
- I2 1700–1800m
- D6 1680–1850m
- A1 1670 m (+L1)
- N5 1600–1800m
- M1 1600m
- S11 1580m
- G8 1560–1580m
- W9 1550–1580m
- G7 1510m
- M2 1500–2100m
- W5 1500–1600m
- W12 1500–1520m
- S12 1500m
- S10 1500m
- S9 1500m
- D4 1500m
- W10 1490–1500m
- I1 1470–1570m
- W11 1450–1600m
- D11 1460m
- D12 1400m
- G6 1400m
- G14 1385 & 1435m
- D1 1370m
- S8 1350m
- G5 1350m
- N11 1300m
- W3 1250m
- D7 1200–1650m
- W13 1200–1400m

LOWER ZONE
- S7 1200m
- D16 1200m
- S6 1100m
- N7 1100m
- G17 1060–1160m
- G13 1060m
- T1 1050–1100 (+W2&M3)
- S5 1050m
- D14 1050m
- D2 1050 (+D8&S4)
- G4 1000–1150m
- D16 1000m
- G3 960–1100m
- G2 940–1160m
- N8 900–1700m
- D3 900950m
- S3 880 m (+N10)
- G16 810–830m
- D13 760m
- N4 700–800m
- S2 700m
- I5 600–650m
- S1 370m

Species	Distribution Type	Occurrences
Equisetum variegatum	5	1
Stellaria longipes s.l.	3	1
Alchemilla glomerulans	6	1
Epilobium sp.	1	1
Gnaphalium norvegicum	6	1
Luzula multiflora	6	1
Phleum commutatum	6	1
Carex macloviana	6	1
Eriophorum scheuchzeri	5	1
Pedicularis flammea	1	1
Antennaria porsildii	4	1
Arnica angustifolia	3	3
Carex lachenalii	6	2
Deschampsia alpina	1	1
Potentilla tridentata	8	1
Diphasiastrum alpinum	6	3
Lycopodium annotinum	6	5
Carex scirpoidea	6	2
Loiseleuria procumbens	6	2
Pedicularis hirsuta	3	2
Gentiana nivalis	6	4
Veronica fruticans	6	6
Phyllodoce caerulea	6	2
Juniperus communis	8	2
Draba aurea	7	2
Veronica alpina	6	4
Cerastium cerastoides	6	2
Diapensia lapponica	6	4
Rumex acetosella	8	1
Gnaphalium supinum	6	3
Tofieldia pusilla	5	5
Bartsia alpina	6	8
Thymus praecox	10	7
Agrostis mertensii	6	4
Dryas integrifolia	1	4
Carex capitata	8	1
Saxifraga paniculata	5	4
Viscaria alpina	6	16
Hieracium alpinum	6	4
Cassiope tetragona	3	5
Woodsia ilvensis	1	6
Huperzia selago	7	7
Erigeron compositus	3	5
Rhododendron lapponicum	3	4
Juncus trifidus	6	12
Empetrum nigrum ssp. hermaphroditum	5	14
Harrimanella hypnoides	6	9

Carex glacialis	3	7												|•|				|•|										|•|																																										
Vaccinium uliginosum	15	5	•	•	•	•	•	•		•	•	•	•		•	•					•									•				•																																											
Carex bigelowii	14	5	•	•		•	•	•		•	•	•	•		•	•					•									•						•																																							
Festuca vivipara	6	6		•								•	•			•			•																			•																																					
Poa alpina	9	6		•		•	•				•	•	•				•				•																			•																																		
Salix glauca	18	5	•	•		•	•	•	•	•	•	•	•		•	•		•	•		•	•																	•		•																																			
Euphrasia frigida	1	5																																										|•|																											
Woodsia glabella	10	3										•	•						•		•						•		•											•				•			•	•																											
Carex norvegica	1	5																																													|•|																								
Ranunculus glacialis	5	4					|•|					|•|							|•|									|•|															|•|																								
Sibbaldia procumbens	6	6					•				•	•	•						•																														•																									
Arabis alpina	3	6											•																																•				•																								
Saxifraga hyperborea	1	1																																														|•|																							
Campanula gieseckiana	27	5		•		•		•	•	•		•	•		•	•		•	•		•	•		•		•	•	•		•	•		•	•			•	•								•	•		•		•																											
Draba subcapitata	1	2																																																	|•|																					
Salix herbacea	18	6	•	•		•	•	•	•		•	•	•		•	•			•						•					•					•	•												•					•																						
Melandrium affine	3	4																																						•				•								•																					
Polygonum viviparum	14	1		•	•	•	•				•	•	•	•	•				•		•						•																					•							•																				
Potentilla crantzii	11	6		•		•			•				•		•									•					•				•		•												•									•																		
Ranunculus pygmaeus	3	1													•																													•													•																
Saxifraga tenuis	6	1				•				•	•																•													•																			•															
Erigeron humilis	24	4				•	•	•	•		•		•		•	•			•					•			•			•			•			•				•			•	•	•		•		•			•		•	•					•													
Campanula uniflora	10	4																	•			•				•					•						•	•	•					•									•							•													
Phippsia algida	1	1																																																											|•|										
Woodsia alpina	5	5										•	•						•																				•																									•											
Chamerion latifolium	27	1		•		•	•		•		•		•		•	•		•			•					•	•	•	•	•			•	•		•	•	•		•					•		•									•	•	•							•										
Cystopteris fragilis	13	8		•							•		•		•	•						•							•	•								•																		•		•					•		•										
Potentilla nivea / hookeriana	19	7									•		•		•	•			•		•	•				•			•				•	•			•	•	•					•				•				•		•										•										
Trisetum spicatum	45	1	•	•	•	•	•	•	•	•	•	•	•	•	•	•	•	•	•		•	•			•	•	•	•	•	•	•	•	•		•	•	•	•	•	•			•	•		•	•					•		•				•	•	•	•										
Poa glauca	26	1			•			•			•		•		•	•			•		•			•		•		•		•			•				•	•	•		•			•			•		•				•			•				•	•	•	•										
Taraxacum sp.	9			•							•	•	•		•		•																								•																	•							•									
Rhodiola rosea	24	6		•		•	•	•	•	•	•	•	•			•			•		•	•								•	•					•		•		•								•		•						•		•		•				•									
Minuartia stricta	2	4																																																											•				•							
Draba norvegica	7	6													•														•									•		•		•																					•							•						
Cardamine bellidifolia	8	2																	•	•							•							•						•						•			•																				•						
Minuartia biflora	12	6				•		•					•		•	•														•				•			•		•									•		•																			•						
Luzula confusa	31	1	•		•	•		•				•				•			•	•	•		•	•		•	•	•	•	•		•		•		•	•		•		•	•		•	•		•	•	•	•		•																			•						
Festuca brachyphylla	6	1																							•												•		•						•				•																				•						
Saxifraga nivalis	34	1			•			•			•		•		•	•		•	•	•	•	•	•	•		•		•	•	•			•	•	•	•	•	•	•				•	•	•		•		•			•	•												•	•				•						
Poa arctica	26	3									•	•		•					•		•		•		•	•		•	•					•	•	•	•		•			•		•	•			•					•		•			•			•						•	•	•				
Luzula spicata	32	6				•	•		•	•	•	•	•			•	•		•	•	•			•	•	•				•				•	•	•		•						•	•		•	•	•						•	•			•		•	•					•			•		
Carex nardina	32	3						•				•	•		•	•		•	•							•	•	•		•							•	•	•					•	•	•	•	•	•				•		•	•		•	•	•	•	•				•			•	•		
Minuartia rubella	12	1						•			•		•						•		•				•			•																	•											•						•								•		•	
Cerastium arcticum	50	1		•	•	•	•	•	•	•	•	•			•	•		•	•	•	•		•	•	•	•		•	•	•			•	•	•	•	•	•	•	•		•	•	•	•		•	•	•	•		•	•		•	•	•	•	•			•	•		•					•	•	
Saxifraga cespitosa	51	1	•		•	•	•	•	•		•	•	•		•	•		•	•	•	•	•				•	•	•	•	•	•		•	•			•	•	•		•	•	•	•	•	•	•	•	•	•	•	•		•	•		•			•		•	•	•		•		•	•	•	•	•	•	
S. oppositifolia	44	1	•		•	•		•	•			•	•		•	•		•	•	•	•		•	•		•	•	•	•	•				•			•	•	•		•		•	•		•	•	•		•		•	•		•		•	•		•	•	•	•						•	
S. cernua	29	1									•		•		•				•					•	•				•	•					•				•		•			•	•	•	•	•		•	•	•	•	•	•				•	•				•	•			•		•	
Draba nivalis	27	1	•										•		•				•	•	•		•			•		•	•			•	•	•		•	•				•	•			•		•		•			•			•	•		•			•							•				•	
Oxyria digyna	39	1		•	•	•	•	•	•	•		•	•			•		•	•	•	•		•	•	•	•	•	•	•	•		•	•	•	•	•		•		•					•			•	•	•						•	•			•		•										•	•	
Erigeron uniflorus	23	6	•	•				•			•	•	•		•	•					•	•				•			•				•	•	•									•	•				•			•			•						•	•											•	
Antennaria canescens	30	6			•	•	•		•	•		•	•		•	•			•			•		•		•			•		•			•			•		•					•	•			•		•		•			•				•		•	•	•								•	•	
Silene acaulis	41	1	•		•		•	•		•	•	•	•		•	•		•	•	•	•			•		•		•		•	•		•	•	•	•	•	•	•	•				•	•		•		•				•		•			•	•			•	•			•		•			•	•
Papaver radicatum	48	1	•								•				•	•		•	•	•	•	•	•	•		•	•	•	•	•	•			•	•	•	•		•		•	•	•	•	•	•	•	•	•			•	•		•		•	•		•	•	•	•	•			•	•	•	•	•	•	•	•
Total number of species			16	35	14	34	29	30	23	15	37	47	52	7	44	31	6	15	37	11	36	14	8	18	9	23	13	17	18	27	8	4	14	20	12	16	22	18	19	11	11	6	6	25	21	5	25	17	17	5	2	15	12	4	19	11	5	9	13	10	18	13	6	2	5	13	5	4	5	7	11	2

Table 2. The altitudinal ranges of species in the southern area, Ammassalik – Kialiip Tasiilaa.

of species of any site (52) and was a rock-bound area of *Empetrum–Vaccinium* heath. The latter site, with only ten species, was a narrow rock rib flanked by glaciers and facing north-west. At about twice the altitude, Stocken's site N13 was on a south-facing slope between 2200–2400 m and had 11 species, while N14 was on north-facing rock ledges at 1900 m and had only six.

It is curious that even common and conspicuous species such as *Silene acaulis* and *Cerastium arcticum* have a more scattered distribution than one would expect. They are certainly capable of growing at all the sites which can support vascular plants and their absence and the general rather random species composition of sites at similar altitudes is perhaps due to the very scattered and fragmentary nature of these montane habitats and reflects the role of chance both in the establishment and extinction of small populations.

To facilitate discussion of the change in species composition with altitude, the sites have been arbitrarily divided into the following three altitudinal zones: a lower zone, 370–1200 m; a middle zone between 1200–1800 m; and an upper zone above 1800 m. The number of species present in each zone is represented diagrammatically in Figure 5a.

5.1.1.1 Lower zone, 370–1200 m

Not surprisingly the species which disappear first are predominantly low arctic and common in the general area of Ammassalik, from the coast to the interior of the fjords. They include dwarf shrubs such as *Juniperus communis, Diapensia lapponica, Phyllodoce coerulea* and *Loiseleuria procumbens,* and herbs such as *Thymus praecox, Bartsia alpina, Veronica fruticans* and *Hieracium alpinum.* However, some species are rather uncommon and occur chiefly towards the head of the fjords where the climate is more continental. Examples are *Cassiope tetragona, Antennaria porsildii, Arnica angustifolia* and *Draba aurea.* It is surprising that the last two species and others such as *Dryas integrifolia* and *Pedicularis hirsuta* should be restricted to this zone since all occur at 910–960 m on the nunataks in Bartholin Land (74° 15′ N), some 900 km further north (Schwarzenbach 1961, although his record is of '*Dryas chamissonis*'). *Carex capitata* was a surprising find since in the Ammassalik area it is known only from two other sites, both near the coast.

The only other site in the Ammassalik area for both the *Antennaria* and *Arnica* is on the slopes of the nearby mountain Cassiopefjeld near the head of Qíngertívaq fjord where they were first discovered by Kruuse (1912). The record of *Potentilla hyparctica* from G4 (Gribbon 1968a and in **STA**) is based on a misidentification of

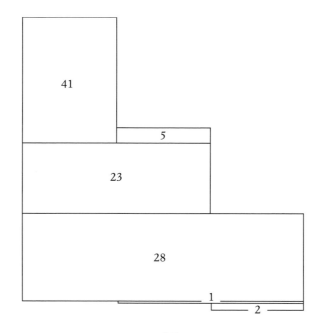
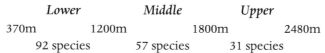
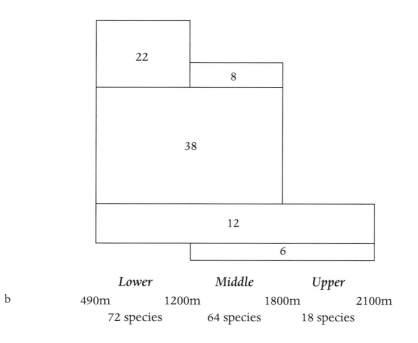

<table>
<tr><td></td><td>Lower</td><td>Middle</td><td>Upper</td></tr>
</table>

Lower   *Middle*   *Upper*

370m   1200m   1800m   2480m

92 species   57 species   31 species

a

Lower   *Middle*   *Upper*

490m   1200m   1800m   2100m

72 species   64 species   18 species

b

Figure 5. The number of species in the three altitudinal zones in a) the southern area and b) the northern area.

P. nivea. Kruuse's record of *P. hyparctica* from Qaqqarsuaq, close to Cassiopefjeld, may also be erroneous and is not represented by any herbarium material.

5.1.1.2 Middle zone, 1200–1800 m

Many predominantly low-arctic species extend into this zone. They are widespread in southern Greenland and have their northern limits in east Greenland, mostly in the central fjord region. Examples include the dwarf shrubs *Salix herbacea*, *Empetrum hermaphroditum* and *Rhododendron lapponicum*, and the herbs *Sibbaldia procumbens* and *Juncus trifidus*. Other species such as *Huperzia selago*, *Campanula gieseckiana*, *Vaccinium uliginosum* and *Carex bigelowii* are widespread over most of Greenland. The low altitudinal limit of *Luzula confusa*, 1650 m, is surprising as this is one of the commonest and most characteristic species of exposed habitats in the high arctic.

Although several species are recorded only from above 1000 m, only three, *Melandrium affine*, *Draba subcapitata* and *Carex glacialis*, can be considered truly montane in the Ammassalik area. The records for the first two are the only ones for the area and they represent appreciable extensions to their known southern limits. Both were found by the Wagers in 1935 and 1936 in the interior of Kangersertuaq (Kangerlussuaq) where *Melandrium affine* occurred as low as 210 m behind Mudderbugt. Although Böcher (1963, p. 74) was obviously aware of their discovery of *Draba subcapitata*, the southern limit of both species is given as Scoresby Sund in Böcher et al. (1978). The only other record in the Ammassalik area for *Carex glacialis* is from the summit plateau (400–500 m) of Nuuk mountain near the north-eastern corner of Sermilik fjord (Böcher 1933, p. 77). Two other species should be included here. *Erigeron compositus* is rare and exclusively montane in south and south-east Greenland, and in the Ammassalik area has a lower altitudinal limit of 550 m. It does not, apparently, come down to sea-level until the north Blosseville Kyst (Halliday et al. 1974). The present records are the only ones for the Ammassalik area although Böcher (1938, p. 189) had correctly surmised that it could well occur 'inland at Ammassalik on the mountain tops'. Its distribution in east and south-east Greenland is shown in Figure 6d.

Several other species appear from Table 2 to have a lower altitudinal limit but this is misleading. *Arabis alpina*, *Ranunculus glacialis*, *R. pygmaeus* and *Euphrasia frigida* are all not infrequent at low altitudes especially towards the coast and *Saxifraga rivularis sensu lato* is quite common. *Carex norvegica* is known in the area only from the nunatak site G7 (Gribbon 1968a) and a subsequent lowland site on the west side of Sermilik (Daniëls & de Molenaar 1970).

5.1.1.3 Upper zone, above 1800 m

Most of the species in this zone have a wide distribution in Greenland and a correspondingly wide ecological amplitude. This is well illustrated by the 11 species in Table 2 which ascend highest. With the exception of *Antennaria canescens* and *Erigeron uniflorus*, both of which are absent from north Greenland, they occur in all the floristic provinces of Greenland (Böcher et al. 1978). They are all common mountain plants and can be regarded as constituting the typical nunatak flora. With few exceptions they occur over the whole altitudinal range up to their individual limits. The commonest species, occurring in more than 40% (19) of the 47 sites in the middle and upper zones, are shown in Table 3.

As in the middle zone there are some species which ascend into the upper zone and have an illusory lower altitudinal limit since they occur elsewhere in the area at low altitudes, for example *Cardamine bellidifolia*, *Draba norvegica*, *Saxifraga cernua*, *Festuca brachyphylla* and *Poa arctica*. Only two species are exclusive to this zone: *Phippsia algida* and *Minuartia stricta*. Like the preceding species, *Phippsia algida* is not specifically montane, being known in the Ammassalik area at sea-level both on the outer coast and in the inner fjords. *Minuartia stricta*, however, is exclusively montane and its occurrence here is remarkable. (It was omitted from my original list in the Report of the Royal Navy Expedition pending verification.) Like *Melandrium affine* and *Draba subcapitata*, its southern limit is given by Böcher et al. 1978 as Scoresby Sund although it too had been found by the Wagers on a nunatak inland from Kangersertuaq. Yet, further north in the central fjord region, *M. stricta* is essentially a lowland plant, its altitudinal limit, as cited by Gelting (1934), being only 600 m.

SPECIES	SITES	SPECIES	SITES
Papaver radicatum	35	*Carex nardina*	24
Saxifraga cespitosa	34	*Saxifraga cernua*	23
Cerastium arcticum	32	*Oxyria digyna*	22
Saxifraga oppositifolia	29	*Saxifraga nivalis*	21
Silene acaulis	26	*Luzula confusa*	20
Trisetum spicatum	24		

Table 3. The commonest species in the middle and upper zones in the southern area.

Two other truly montane species should be mentioned here: *Potentilla nivea* (lower limit 900 m) and *Papaver radicatum* (360 m). Prior to these investigations both were known in the area only from Kruuse's (1912) records from Cassiopefjeld where they were found at c. 600 m. They occurred at 18 and 47 sites respectively out of the 70 montane sites considered here; in fact *Papaver radicatum* is one of the commonest montane species. Additional records were made by F.J.A. Daniëls (in litt.) from east of Kuummiut and Kialiip Tasiilaa in 1969 and 1966 respectively. Further records for *Potentilla nivea* were contributed by L. Kliim-Nielsen who found it in 1969 at 350 m on Qaqqarsuaq, a mountain on the opposite side of Qíngertivaq to Cassiopefjeld, and by C.L. Bagley in 1985 at 1300 m on a mountain 10 km south-west of Cassiopefjeld.

Table 4 gives a summary of those species which appear to be exclusively montane in the Ammassalik area. All are most frequent in east Greenland in the central fjord region.

As one would expect, Table 2 demonstrates very clearly that the low-arctic species tend to disappear first, that the widespread often circumpolar species ascend highest. In general, the higher the latitudinal limit of a species the higher it ascends. Conversely, high-arctic species with a sea-level southern limit in east Greenland far to the north of Ammassalik may occur substantially further south on the nunataks, and the further south, the higher the altitude and the further inland. It is for these reasons that the existence in south-east Greenland of *Minuartia stricta*, *Melandrium affine* and *Draba subcapitata* went for so long undetected.

SPECIES	ALTITUDE	SPECIES	ALTITUDE
†*Papaver radicatum*	370–2450 m	*Campanula uniflora*	?–1780 m
*†*Draba subcapitata*	1600 m	†*Erigeron compositus*	880–1300 m
*†*Minuartia stricta* (two sites)	2000 m; 1800–2480 m	*Antennaria porsildii*	400–1500 m
*†*Melandrium affine*	1500–1670 m	*Carex glacialis*	400–1400 m
†*Potentilla nivea*	350–1900 m		

Table 4. Highest recorded altitudes of exclusively montane species in the southern area.

* southern limit in east Greenland
† south-east Greenland distribution maps shown in Figure 6 and Figure 8.

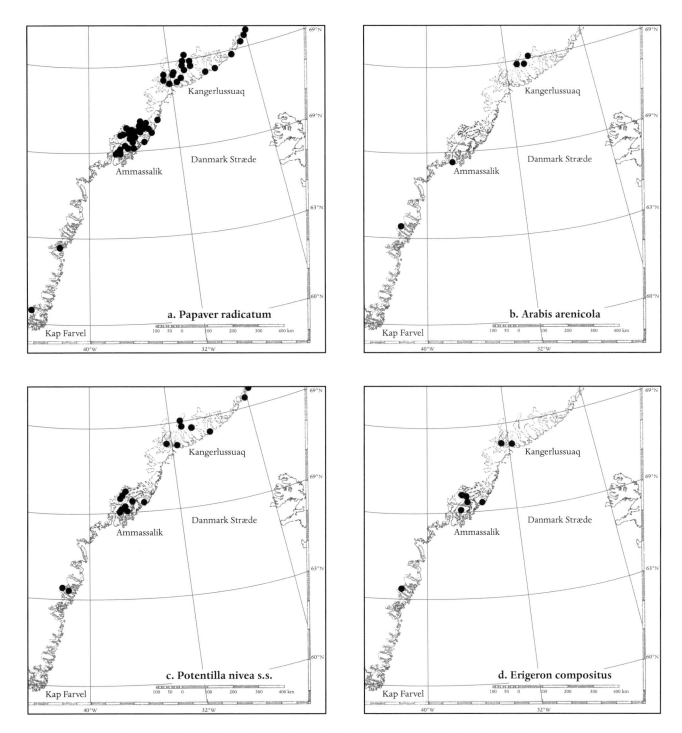

Figure 6. The south-east Greenland distributions of four predominantly medium-
and high-arctic species:

a) *Papaver radicatum*, b) *Arabis arenicola*, c) *Potentilla nivea* and d) *Erigeron compositus*.

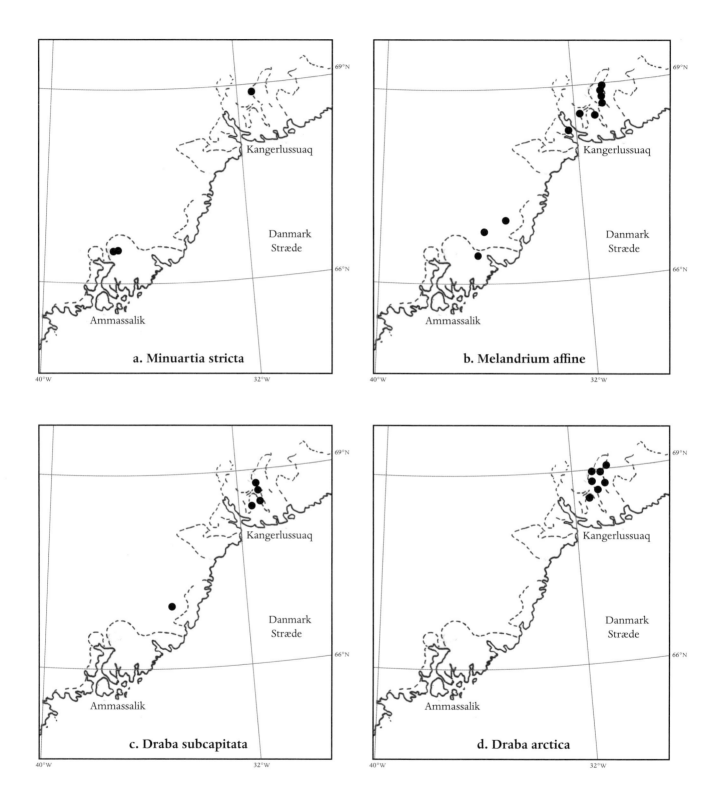

a. **Minuartia stricta**

b. **Melandrium affine**

c. **Draba subcapitata**

d. **Draba arctica**

ANALYSIS OF THE PLANT RECORDS

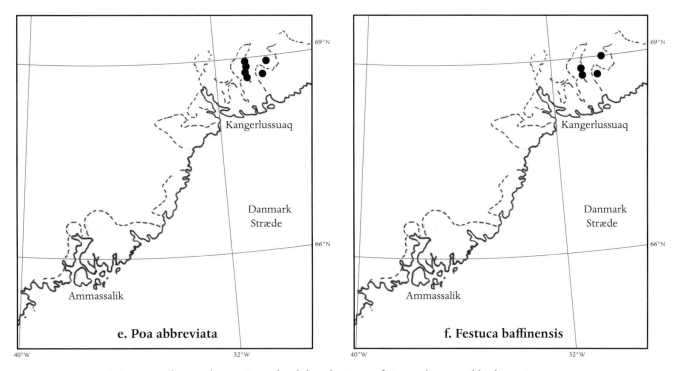

Figure 7. The south-east Greenland distributions of six medium- and high-arctic species: a) *Minuartia stricta*, b) *Melandrium affine*, c) *Draba subcapitata*, d) *Draba arctica*, e) *Poa abbreviata* and f) *Festuca baffinensis*, all of which are here at their southern limits in east Greenland.

5.1.2 The rate of decrease of the flora with altitude

This was studied by Gribbon (1968a) and earlier in his Report of the 1963 Scottish East Greenland Expedition using his data from the 'Caledonian Alps', inland from Ammassalik. He concluded that for the most favourable sites the decrease in the number of species above 1000 m was fairly constant at about five per 100 m. For less favourable sites the decrease was appreciably less, about one species per 100 m.

It is not possible to repeat this exercise using all the data in Table 2, as the necessary habitat information is inadequate for most of the sites. Instead the number of species at each of 18 selected sites has been plotted against altitude (Figure 8a); these sites are those with the maximum number of species at altitudinal intervals of approximately 100 m. A regression analysis of this data shows that the decrease is only 2.1 species per 100 m. This is less than half

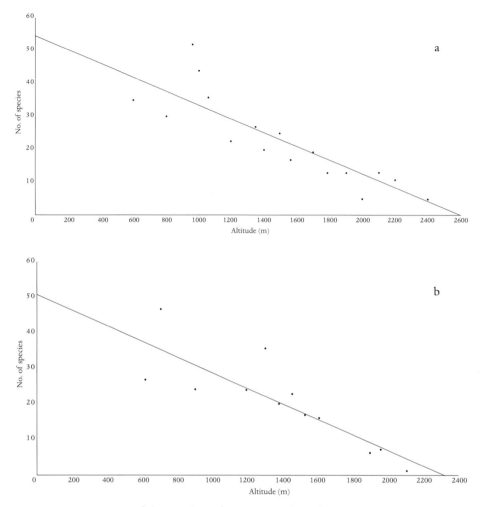

Figure 8. Comparison of the number of species at selected species-rich sites plotted against altitude in a) the southern area and b) the northern area, and their respective regression lines (correlation coefficients -0.74 and -0.71 respectively).

Gribbon's figure and is no doubt due to the fact that the present study extends his altitudinal range by 900 m. One effect of this will be to reduce the effect of his very rich site G3 (52 species) on the rest of the data.

Applying a regression analysis to Gribbon's data for his seven 'richest' sites produces an upper limit for vascular plants of only about 1950 m; the comparable figure from Figure 8a is 2550 m. Ten sites in Table 2 are wholly or partly from 2000 m or above. Two of these, N12 (1800–2480 m) and N13 (2200–2400 m), in Schweizerland, were visited by Stocken on the 1966 Naval Expedition and at each he noted the maximum altitudes at which individual species were seen. These are given in Table 5 for the 15 species which occurred above 2200 m.

The discrepancy between Gribbon's and Stocken's results is probably attributable to the fact that the mountains visited by Gribbon are both more coastal and lower than those in Schweizerland, and both these factors can operate to produce a lower altitudinal limit. Yet Stocken's figure is too high to be considered typical. There is evidence that north of Femstjernen the limit falls: apart from Stocken's site N13 and the Westminster site W7 the number of species recorded from the six sites at or above 2000 m does not exceed six and at the highest, L3 (2300 m) by Kristian Gletsjer, only two species were found. Although Porter, of the 1968 Army Expedition, investigated ten possible sites between 2075 m and 2450 m north of this glacier and east of Mont Forel, he failed to find any vascular plants. Quite remarkable is the large number of species, 13, recorded at the highest and most inland site, W7 (2100–2150 m), visited by the 1974 Westminster Expedition slightly further to the north-east.

All one can say is that over most of the nunatak area the limit lies between 2150 and 2480 m but is appreciably lower nearer the fjords. Stocken, who was killed a few days after making his observations at 2480 m, never knew he had the distinction of having extended the known altitudinal limit for vascular plants not only in Greenland but in the Arctic as a whole.

SPECIES	ALTITUDE	SPECIES	ALTITUDE
Minuartia rubella	2480 m	*Antennaria canescens*	2300 m
Papaver radicatum	2450 m	*Cerastium arcticum*	2300 m
Saxifraga cernua	2450 m	*Draba nivalis*	2300 m
S. oppositifolia	2400 m	*Oxyria digyna*	2280 m
S. cespitosa	2400 m	*Silene acaulis*	2200 m *
Carex nardina	2330 m	*Erigeron uniflorus*	2200 m *
Poa glauca	2330 m	*Luzula spicata*	2200 m *
Phippsia algida	2330 m		

Table 5. Maximum altitudes of species recorded by Stocken above 2200 m. (*Minuartia stricta* (N12, 1800–2480 m) should be included here but it was not recognised in the field and its maximum altitude is therefore not known.)

* 2300 m at site L3

5.1.3   The effect of altitude on the biological distribution types represented in the flora

Böcher et al. (1959) divided the Greenland flora into ten biological distribution types (Table 6) based primarily on climates. Their allocation of species was subsequently slightly modified by Böcher (1963) who also added one more type. This was further refined by Bay (1992) who recognised 14 types. In order to allow comparison of the present data with those of Gribbon (1968a and b) and Halliday (1967) Böcher's 1959 analysis is used here. These geographical elements form a convenient and instructive way of analysing quantitatively the floristic changes with altitude, a method which is complementary to the consideration of individual species in the first part of this section. The first person to analyse montane data in this way was Gribbon (1968a) using his own expedition data from 17 sites in Schweitzerland. His main conclusions were that with increasing height the percentage of low-arctic oceanic montane species (Type 6) decreased dramatically. This was paralleled by the decrease in the smaller group of widespread low- or medium-arctic species (5). Corresponding with these declines is the increasing percentage of the widespread arctic montane species (1) which dominate the flora at higher altitudes. He also noted that the two small groups of widespread arctic continental species (3) and medium-arctic montane species (4) were appreciably commoner between 1000 and 1600 m. He concluded by likening an increase of altitude to an increase of latitude or continentality. My analysis of Stocken's data in the Report of the Royal Naval East Greenland Expedition 1966 (Halliday 1967), also from Schweizerland, showed essentially the same features although Type 5 was appreciably smaller, no doubt reflecting the greater continentality of Stocken's inland sites.

Gribbon also found that, at the same altitude, sites with a poor flora had a higher percentage of Type 1 species and a lower percentage of Type 6 species than species-rich sites. The distinction as to what are favourable and unfavourable sites can really only be made in the field and no attempt will be made to make such a comparison on the data in Table 2.

The only types with more than ten species (i.e. 10%) in Table 2 are 1, 5 and 6. Using the same 18 sites ranging from 600–2400 m as in Figure 8a, the changing percentages with altitude of these three types are shown in Figure 9a. Again these demonstrate the increasing importance of Type 1 species with increasing altitude and the decline in Types 5 and 6. As with Stockens data, the decline in Type 6 is less striking than that found by Gribbon and this is no doubt due

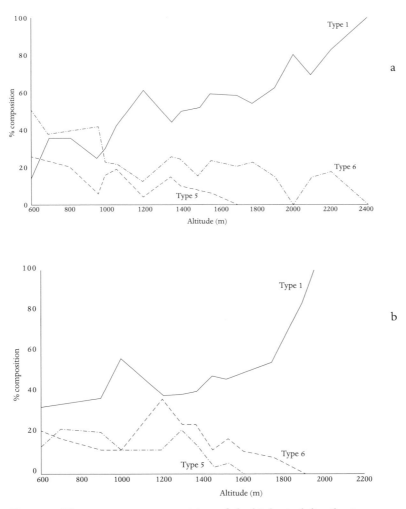

Figure 9. The percentage composition of the biological distribution types 1 (arctic montane), 5 (low- or medium-arctic) and 6 (low-arctic oceanic) species in a) the southern area and b) the northern area.

to the greater altitudinal range of the present data and the remarkable ability of certain Type 6 species, notably *Erigeron uniflorus*, *Antennaria canescens* and *Luzula spicata*, to ascend to over 2000 m.

These general conclusions can be amplified by Table 6, which shows the distribution of all the distribution types in the three altitudinal zones referred to earlier in this section.

BIOLOGICAL DISTRIBUTION TYPE	1	2	3	4	5	6	7	8	9	10	NC	TOTAL
Lower zone, 360–1200m	21	-	9	4	12	35	3	5	-	11	2	92
Middle zone, 1200–1800m	21	1	5	4	8	14	2	1	-	-	1	57
Upper zone, 1800–2400m	18	-	2	1	1	6	2	1	-	-	1	31
TOTAL	24	1	9	6	14	35	3	5	-	1	2	100

Table 6. The number of species of each biological distribution type in the three altitudinal zones.

1 Widespread arctic montane
2 High-arctic
3 Widespread arctic continental
4 Medium-arctic montane
5 Widespread low- or medium-arctic
6 Low-arctic oceanic montane
7 Low-arctic continental
8 Widespread boreal
9 Boreal or low-arctic silvicolous
10 Boreal sub-oceanic
NC Not classified

As mentioned above, Gribbon noted peaks in the frequencies of Type 3 and 4 species between 1000 and 1600 m. There is little evidence of this in the present data. There are nine Type 3 species in Table 2 and their altitudinal limits vary widely. *Arnica angustifolia*, *Pedicularis hirsuta* and *Stellaria longipes* are restricted to the lower zone but *Poa arctica* and *Carex nardina* ascend to over 2000 m. Of the six Type 4 species, *Ranunculus glacialis* and *Erigeron humilis* are recorded only from the lower zone, *Melandrium affine* only from the middle and *Minuartia stricta* only from the upper.

If the somewhat speculative analysis of the montane climate (pp.. 17–18) is correct, a climatic gradient should exist from the head of the fjords to the interior. In theory this change should occur at the same altitude but because the low ground in the interior is largely ice-covered, any changes observed ascending the glaciers will be obscured by change in altitude. Gribbon (1968a) examined his data from the inland mountains in the south of the area and concluded that there was no evidence of any such climatic difference. It would be surprising, however, if this were true over the area as a whole but the complex geography means that any gradient is unlikely to be a simple one. For example, the two large fjords, Sermilik and Kangersertivattiaq (Kangertítivatsiaq), which delimit the southern part of the area on the west and east respectively, are likely to exert an ameliorating effect on the climate far up the Nigertiip

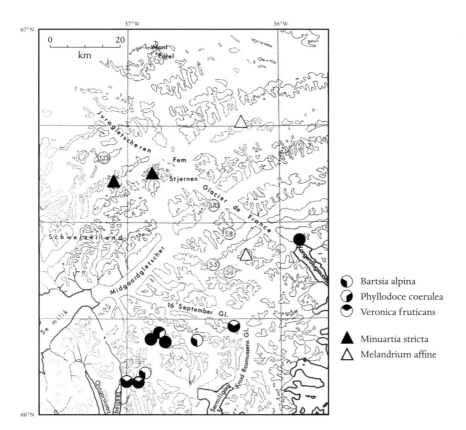

Figure 10. The contrasting distributions in the southern area of the low-arctic oceanic species *Phyllodoce caerulea*, *Bartsia alpina* and *Veronica fruticans* (biological distribution type 6), and the medium-arctic *Minuartia stricta* and *Melandrium affine* (type 4).

Apusiia ('Midgårdgletsjer') and Kattilersorpia ('Glacier de France'), the broad Sermilik fjord having the greater effect.

This general picture of an increasingly continental climate inland from the heads of the fjords is well illustrated by Figure 10 showing the complementary montane distributions of three Type 6 and two Type 4 species.

The climatic analysis also suggests that the area might include representatives of the following:

1 Coastal species which become increasingly rare inland, and being intolerant of both low temperatures and low humidity, are restricted to the fjords or lower slopes.

2   Outer coastal species which inland become montane, requiring humid conditions and cool summers.

3   Mid- to high-arctic species which are largely restricted to the inland mountains, being intolerant of both the warm summers around the heads of the fjords and the high humidity of the outer coast.

The two contrasting sets of species mapped in Figure 10 illustrate the first and last of these categories. The second is represented only by *Ranuculus glacialis* and then not very convincingly.

5.2   The northern area: Kangersertuaq (Figure 4, p. 34; Table 7) and comparison with the southern area

Kangersertuaq has an appreciably smaller flora than the Ammassalik area. This is not surprising since not only is it further north and a substantially smaller area but proportionally more of the low-lying land is ice-covered. 86 species, representing 63% of the flora of Kangersertuaq, are listed in Table 2. This is unexpectedly high since only half as many sites were visited as in the southern area.

5.2.1   The altitudinal ranges and distribution of species

Using the same altitudinal zones as applied to the southern area, the distribution of species between them is shown in Figure 5b (p. 45).

5.2.1.1 Lower zone, 490–1200 m

Comparison with Figure 5a shows a decrease in the proportion of species in the lower zone from 92% to 84% but, more significantly, far fewer species are restricted to this zone, 25% compared with 41%. In view of its more northerly situation it might have been expected that there would be a general lowering of altitudinal limits in the Kangersertuaq area and with it an increase, not a decrease, in the proportion of species restricted to the lower zone.

However, of the 41 species exclusive to this zone in the south:

6 are absent from the Kangersertuaq area,
20 are present but are absent from the mountains of the inner fjord,
9 are present and restricted to the lower zone,
6 are present and extend into the middle zone.

The absence from Table 7 of so many species which are lowland at Ammassalik suggests that their altitudinal limits are very much lower here, and certainly the seven species which extend to the middle zone in the Ammassalik area but are restricted to the lower zone in the north show reduced altitudinal limits ranging from c. 450 m (*Rhododendron lapponicum*) to 1000 m (*Ranunculus pygmaeus*). On the other hand, the species which are restricted to the lower zone in both areas have rather similar altitudinal limits and the six which actually extend into the middle zone at Kangersertuaq, for example *Carex lachenalii*, *C. scirpoidea* and *Cassiope tetragona*, have limits which are 100–300 m higher. It seems likely therefore that it is the drier climate of the inner part of the fjord and the interior which is responsible for the absence from this zone of so many of the Ammassalik species.

Böcher (1933a, b), who visited the fjord in 1932 and 1933, was impressed by the contrast between the impoverished flora of the inner fjord and the richness of the outer parts. He attributed this to what he regarded as the inimical climate of the interior characterised by drought and cold winds and he went on to express his belief (1933a, p. 27) that the climate there is not much warmer than that at the mouth of the fjord. He used the distribution of *Cassiope tetragona*, which is absent from the outer fjord, to delimit the two areas. Of the 20 species referred to above which are lowland in Kangersertuaq, 11 are restricted to the outer parts of the fjord and penetrate no further than Kraemer Ø. These include *Alchemilla glomerulans*, *Potentilla tridentata*, *Loiseleuria procumbens*, *Thymus praecox* (Elsley & Halliday 1971, p. 14) and *Phleum commutatum*.

However, our knowledge of the inner fjord is still rather limited. Three sites have each produced only 19 species and one 13. By contrast, the general Mudderbugt area, about half-way up the fjord, has 82 species (Devold & Scholander 1933) and overlooking the very head of the fjord is site K1 (Table 7)

Species legend (columns 1–40):

1 Erigeron compositus · 2 Equisetum arvense · 3 E. variegatum · 4 Sedum villosum · 5 S. annuum · 6 Draba aurea · 7 Stellaria longipes s.l. · 8 Gentiana nivalis · 9 Veronica alpina · 10 Euphrasia frigida · 11 Tofieldia pusilla · 12 Arnica angustifolia · 13 Rhododendron lapponicum · 14 Juncus trifidus · 15 Ranunculus pygmaeus · 16 Carex capillaris · 17 Woodsia ilvensis · 18 Carex glacialis · 19 Sagina intermedia · 20 Potentilla crantzii · 21 Veronica fruticans · 22 Viscaria alpina · 23 Cerastium cerastoides · 24 Phyllodoce caerulea · 25 Carex bigelowii · 26 Poa alpina · 27 Saxifraga rivularis · 28 Woodsia alpina · 29 Poa arctica · 30 Carex lachenalii · 31 Saxifraga paniculata · 32 Empetrum nigrum · 33 ssp. hermaphroditum · 34 Harrimanella hypnoides · 35 Carex scirpoidea · 36 C. norvegica · 37 C. misandra · 38 Minuartia biflora · 39 Salix herbacea · 40 Huperzia selago

Zone	Site (elevation)	1	2	3	4	5	6	7	8	9	10	11	12	13	14	15	16	17	18	19	20	21	22	23	24	25	26	27	28	29	30	31	32	33	34	35	36	37	38	39	40
UPPER ZONE	B2b 2100m																																								
	L1 2000m																																								
	W14 2000m																																								
	C1 1950m																																								
	B2a 1950m																																								
	A3 1950m																																								
	A2 1890m																																								
	W13 1830m																																								
MIDDLE ZONE	W15 1760–1900m																																								
	C3 1750m																																								
	A4 1740m																																								
	W12 1600m																																								
	W8 1600m																																								
	W6 1550m																																								
	A6 1520–1710m																																								
	C2 1500m																																								
	W11 1500m																																								
	A5 1460–1740m																																								
	W39 1450m																																								
	W5 1450m																																								
	W10 1440m																																								
	W7 1400–1420m																																								
	W17 1400m																																								
	A7 1370m																																								
	B1 1300–1400m																										•			•	•	•	•	•	•	•	•	•	•	•	•
	W3 1300m																											•													
	W40+C4 1300m																									•															
	W38 1300m																																								
	A10 1280–1400m																													•				•					•		
	W16 1200–1400m																													•											
	W37 1200–1300m																								•	•															
	W1 1200–1300m																								•				•												
LOWER ZONE	A8 1190–1310m																				•	•	•				•			•											
	B3 1000m																			•	•						•								•	•			•		
	S1 900–1060m																						•							•	•			•							
	A14 850–1220m																		•																						
	A11 790m																					•	•																•		
	W27 760m																•													•						•				•	
	K1 700–1000m		•	•	•	•	•	•	•	•	•	•	•		•	•					•		•			•	•					•	•			•	•				
	W34 610m	•											•	•	•																		•								•
	A9 490–940m														•								•																		

	Species	Erigeron compositus	Equisetum arvense	E. variegatum	Sedum villosum	S. annuum	Draba aurea	Stellaria longipes s.l.	Gentiana nivalis	Veronica alpina	Euphrasia frigida	Tofieldia pusilla	Arnica angustifolia	Rhododendron lapponicum	Juncus trifidus	Ranunculus pygmaeus	Carex capillaris	Woodsia ilvensis	Carex glacialis	Sagina intermedia	Potentilla crantzii	Veronica fruticans	Viscaria alpina	Cerastium cerastoides	Phyllodoce caerulea	Carex bigelowii	Poa alpina	Saxifraga rivularis	Woodsia alpina	Poa arctica	Carex lachenalii	Saxifraga paniculata	Empetrum nigrum	ssp. hermaphroditum	Harrimanella hypnoides	Carex scirpoidea	C. norvegica	C. misandra	Minuartia biflora	Salix herbacea	Huperzia selago
	Distribution Type*	3	8	5	8	6	7	5	6	6	5	5	3	3	6	5	6	5	7	1	6	6	6	6	6	5	6	5	5	3	6	6	5	6	5	5	3	5	6	8	8
	Occurrences	1	1	1	1	1	1	1	1	1	1	2	3	2	1	1	1	1	1	2	1	1	4	2	1	2	5	3	1	7	3	3	6	5	3	1	1	4	2	2	2

Species		
Festuca brachyphylla	7	5
Vaccinium uliginosum	7	5
Sibbaldia procumbens	4	6
Polygonum viviparum	5	1
Cardamine bellidifolia	4	1
Cassiope tetragona	7	3
Saxifraga tenuis	8	1
Minuartia rubella	6	1
Hierochloe alpina	3	3
Potentilla hyparctica	3	2
Arabis alpina	3	6
Campanula gieseckiana	8	5
Taraxacum sp.	3	
Draba nivalis	4	1
Luzula confusa	13	1
Salix glauca	14	5
Erigeron humilis	4	4
Rhodiola rosea	6	6
Luzula spicata	8	6
Draba subcapitata	5	2
Phippsia algida	5	1
Campanula uniflora	8	4
Potentilla nivea /hookeriana	6	7
Antennaria canescens	11	6
Silene acaulis	15	1
Oxyria digyna	15	1
Carex maritima	4	3
Melandrium affine	10	4
Minuartia stricta	2	4
Draba norvegica	11	6
Poa abbreviata	8	2
Carex nardina	18	3
Chamerion latifolium	10	1
Saxifraga nivalis	19	1
Trisetum spicatum	20	1
Saxifraga oppositifolia	18	1
Draba arctica	8	3
Erigeron uniflorus	7	6
Cerastium arcticum	28	1
Arabis arenicola	3	4
Festuca baffinensis	4	2
Cystopteris fragilis	11	8
Papaver radicatum	18	1
Saxifraga cernua	20	1
Saxifraga cespitosa	26	1
Poa glauca	23	1
Ranunculus glacialis	3	4
Total number of species	16	27

Table 7. The altitudinal ranges of species in the northern area, Kangersertuaq.

at 700–1000 m on Batbjerg, visited by C.K. Brooks in 1975 and by far the richest of the Kangersertuaq montane sites with 47 species. This is only five less than the richest site in the southern area and it actually includes five species listed by Böcher (1933a, p. 26) as characteristic of the outer coast. Certainly, conditions here and at Mudderbugt are very favourable for the more thermophilous and continental species and Brooks (1983 in litt.) has described the vegetation at Batbjerg as "really verdant, much more so than by the coast".

The evidence from Brooks' site, Table 7 as a whole, and the Norwegian lists from the Mudderbugt area certainly suggests that there is no general impoverishment of the flora in the central and inner parts of the fjords, a conclusion in agreement with the recollections of expedition members, referred to earlier, on the very favourable climate of the interior.

Two species recorded only from the Batbjerg site are new to Kangersertuaq: *Draba aurea* and *Stellaria longipes* s.l. The former record fills a gap between Scoresby Sund and Kap Warming (67° N) and the latter one between Turner Sund (Halliday et al. 1974) and the Kattilersorpia site (Table 2: I5) above the head of Kangersertivattiaq.

The six species which are absent from the Kangersertuaq area comprise southern species which do not extend this far north: *Juniperus communis*, *Gnaphalium norvegicum* and *Deschampsia alpina*, and others which, surprisingly, have here a gap in their east coast ranges: *Dryas integrifolia* (*D. octopetala* is also absent), *Antennaria porsildii* and *Rumex acetosella*.

The 22 species exclusive to this zone consist of the nine referred to above which were also present in this zone in the southern area; five species which although present in the Ammassalik area are unrepresented in Table 2: *Equisetum arvense*, *Sagina intermedia*, *Sedum annuum*, *S. villosum* and *Carex capillaris*; and seven which, in the southern area, extended into the middle zone. In addition to *Rhododendron lapponicum* and *Ranunculus pygmaeus* which have already been mentioned, these include *Carex glacialis* and *Potentilla crantzii*.

5.2.1.2 Middle zone, 1200–1800 m

Proportionally more species occur in this zone in the Kangersertuaq area (75%) than in the southern area (57%). As mentioned above, six species have their upper altitudinal limits in this zone compared with the lower zone in the south but 19 species with their limits in the upper zone in the south have moved down into the middle zone. The latter group includes the two which in the south were exclusive to the upper zone, *Minuartia stricta* and *Phippsia algida*, and also *Luzula spicata* and *L. confusa*, the latter again having the lower altitudinal limit of the two.

The most interesting species are the seven which are unrepresented in the Ammassalik list (Table 2). *Hierochloe alpina* is absent between Kangersertuaq and the Skjoldungen area. *Carex maritima* has a very disjunct distribution in east Greenland south of Scoresby Sund. The Kangersertuaq localities, both by the fjord and on the nunataks, are shown on Böcher's (1954, p. 239) map and to this should be added the present records from Ammassalik and Kap Russel (69° 59′ N) (Halliday et al. 1974). Its occurrence on nunataks at Kangersertuaq is not unexpected since Gelting (1934), for example, mentioned that it was common between 800–1000 m on Clavering Ø (74° 15′ N) and Schwarzenbach (1961) recorded it at 910 m on nunataks to the west of this. It has also been reported at 1050 m and 1240 m in two nunatak areas of south-west Greenland (Frederiksen 1971, Young 1980).

The high-arctic sedge *Carex misandra* is known in the Kangersertuaq area only from the Mittivakkat nunatak (site B1), from near Mudderbugt (Devold & Scholander 1933), and from above the head of Watkins Fjord and by the Schelderup Gletsjer in 2-miippugut (J.C. Jacobsen Fjord) where it was discovered by the 1972 Westminster Expedition and the Wagers respectively. Its southern limit on the east coast was extended from here to the Ammassalik area by L. Kliim-Nielsen who found it at Paarnagajîit, on the east side of Sermilik, in 1971. In view of this last find it is very curious that it was not found at any of the sites in Table 2. Another high-arctic species is *Potentilla hyparctica* which has its southern limit in Kangersertuaq where it is known only from three nunatak sites.

Arabis arenicola is a very rare species in east Greenland with a curiously patchy distribution. Until recently it was known only from two sites, in Scoresby Land and in Jameson Land. Its known range has now been considerably extended (Figure 6b): northwards to Traill Ø and Lindeman Fjord (74° 39′ N) where it was found respectively by H.A. Lang in 1991 and by G. Halliday in 1980, and southwards at the three nunatak sites in Table 7, at lowland sites at Ammassalik (Daniëls & de Molenaar 1970) and more recently at Skjoldungen (63° 28′ N) by Bay (1993). It has also been reported from nunataks inland from Frederikshåb, south-west Greenland, at altitudes of between 1200–1400 m (Frederiksen 1971).

In the discussion of Table 2 it was pointed out that *Draba subcapitata*, *Minuartia stricta* and *Melandrium affine*, all previously thought to have their southern limits at Scoresby Sund (Böcher et al. 1978), all occur far to the south on nunataks both in the Ammassalik and Kangerlussuaq areas. On the latter nunataks they are joined by three more species also previously unknown south of Scoresby Sund: *Draba arctica*, *Festuca baffinensis* and *Poa abbreviata*. Although descending to sea-level further north in the central fjord region, *Poa abbreviata* is essentially a plant of the high mountains. It was one of the commonest plants observed by Schwarzenbach (1961) on nunataks west of Clavering Ø and ascended highest, to about 1650 m. Distribution maps of these six species within the study area are shown in Figure 7. These records were incorporated by Bay (1992) in his maps of total Greenland distributions.

5.2.1.3 Upper zone, above 1800 m

The number of species in this zone is far fewer than in the southern area, 16 compared with 33. *Arabis arenicola*, *Festuca baffinensis* and *Poa abbreviata* are the only ones unrepresented in the Ammassalik list. As noted above, the upper limits of 19 species have moved down to the middle zone and these include *Minuartia rubella*, *M. stricta*, *Silene acaulis*, *Oxyria digyna*, *Antennaria canescens*, *Erigeron uniflorus*, *Luzula spicata* and *Phippsia algida*, exactly half of the species listed in Table 5 (p. 53) as having the highest altitudinal limits in the southern area. The 16 species in this zone are listed in Table 8 with their highest recorded altitudes for comparison.

The presence and position in this list of *Ranunculus glacialis* is at first sight surprising since this is predominantly an outer coastal species, although at its southern limit, Ammassalik, and at Kangersertuaq it penetrates some distance into the fjords. Other montane records are from Clavering Ø (Gelting 1934, 1200 m) and the high nunataks to the west (Schwarzenbach 1961).

5.2.2 The rate of decrease of the flora with altitude

Using, as before, sites selected for their floristic richness, the relationship between the number of species and altitude is shown in Figure 8b. The regression line shows an almost identical rate of decrease with altitude in the two areas, 2.1 species per 100 m in the south compared with 2.2 species in the north, and suggests an altitudinal limit of about 2400 m. This compares with a figure of 2550 m in the southern area. Both figures are actually somewhat higher than those actually observed: the highest site in the southern area being

SPECIES	ALTITUDE	SPECIES	ALTITUDE
Papaver radicatum	2500 m	*Festuca baffinensis*	1950 m
Poa glauca	2500 m	*Saxifraga oppositifolia*	1950 m
Saxifraga cernua	2500 m	*Trisetum spicatum*	1950 m
S. cespitosa	2500 m	*Saxifraga nivalis*	1890 m
Ranunculus glacialis	2100 m	*Chamerion latifolium*	1890 m
Cystopteris fragili	2040 m	*Carex nardina*	1830 m
Arabis arenicola	1950 m	*Draba arctica*	1830 m
Cerastium arcticum	1950 m	*Poa abbreviata*	1830 m

Table 8. The altitudinal limits of species occurring in the upper zone of the northern area.

2480 m and in the northern 2100 m. If, as seems likely, the montane climate in the north is not only colder but more continental, this would account for the virtual disappearance of the low-arctic oceanic montane species (Type 6) from the upper zone.

The near linear decrease in the flora in both areas might appear to be at variance with the results of Bruun et al. (2006) and others. Bruun's data were based on transects upwards from the tree-line in northern Scandinavia. These showed an early peak followed by a long near linear decline. Such a peak in Greenland would occur below the lowest altitudes in the present study.

5.2.3 The effect of altitude on the biological distribution types represented in the flora

The percentage frequency of widespread arctic montane species (Type 1), low- or medium-arctic species (5) and low-arctic oceanic montane species (6) at the 14 selected sites are shown in Table 9. As in the Ammassalik area, Type 1 species play an increasingly important role with increasing altitude and eventually dominate the flora. In fact seven out of the eight species occurring above 1900 m belong to this group. The decrease in Types 5 and 6 is again generally similar to that in the south although both play a less prominent

BIOLOGICAL DISTRIBUTION TYPE	1	2	3	4	5	6	7	8	9	10	NC	TOTAL
Lower zone, 490–1200 m	19	1	8	3	15	19	2	3	-	-	1	71
Middle zone, 1200–1800 m	18	4	7	6	11	14	1	2	-	-	1	64
Upper zone, 1800–2100 m	9	2	1	2	-	1	-	1	-	-	-	16
TOTAL	19	4	10	6	17	22	3	4	-	-	1	86

Table 9. The number of species of each biological distribution type in the three altitudinal zones in the northern area (see Table 6, p. 56, for distribution types).

role in the north and, as mentioned above, Type 6 is absent from the highest sites, only one species, *Draba norvegica*, being present in the upper zone. The presence of nine Type 6 species at site A8 at 1190 m high up on the east side of the middle part of the Rosenborg Gletsjer is surprising; one of them, *Veronica fruticans*, is represented in Table 7 only from this site.

The frequency of all the biological distribution types is shown in Table 9.

Most of the differences between this table and Table 6 from the southern area have already been commented on. The most striking is the decrease in the number of Type 6 species in the north, the Ammassalik area having twice as many such species in the lower zone. On the other hand, Types 2 and 5 are slightly better represented, indicating, perhaps, a more rigorous climate.

Chapter 6
Comparison with other areas of Greenland

6.1   West Greenland

Subsequent to his 1963 east Greenland expedition, Gribbon led a number of climbing expeditions to west Greenland, visiting the area south of Evighedsfjord in 1965 and Upernivik Ø, near Uummanaq, in 1967 and 1969. On all of these he continued his work on the altitudinal ranges of vascular plants, with the results appearing in his various expedition reports (Gribbon 1965, 1967, 1969). He found few differences between his data from east and west Greenland. There was the same increase with altitude of the widespread arctic species (Type 1) and a decrease in the low-arctic montane element (6). At Evighedsfjord (64° 50′ N) the mountain flora was appreciably poorer than at Ammassalik but on Upernivik Ø (71° 15′ N) the arctic-continental (3) and medium-arctic species (4) were better represented than at either of the other two areas, the former type comprising 22% of the flora up to 1000 m and the latter reaching 18% at 1400 m.

He noted a similar decrease with altitude in the number of species in each of the three areas, about five per 100 m. As mentioned earlier, his Ammassalik figure may be too high, and plotting his data for Evighedsfjord and Upernivik Ø on the same basis as earlier in this paper, using selected species-rich sites and omitting ones below 300 m, regression lines for the two areas give rates of decrease of 4.3 for Evighedsfjord and 3.2 for Upernivik Ø, and altitudinal limits of 1680 m and 2300 m respectively (see Table 10), the latter figure exceeding the highest mountains on the island. There was actually one species, *Papaver radicatum*, on the summit of the highest mountain south of Evighedsfjord, 1680 m, coinciding with the calculated maximum height for vascular plants. The Upernivik Ø figures need to be treated with some caution since only two of the sites were above 1550 m. Nevertheless it is probably true that in this area with its more continental climate the altitudinal limit is significantly higher than at

	ALTITUDINAL RANGE	RATE OF DE-CREASE IN SPE-CIES PER 100 M	CORRELATION COEFFICIENT	MAXIMUM ALTITUDE	
				CALCULATED	OBSERVED
S. Greenland, 60°15′ N Kap Farvel Gribbon (1971)	730–1370 m	6.6	-0.72	1500 m	1370 m
W. Greenland, 64°50′ N Evighedsfjord Gribbon (1965)	390–1680 m	4.3	-0.81	1680 m	1680 m***
W. Greenland, 71°15′ N Upernivik Ø Gribbon (1967 & 1983 in litt.)	360–1850 m	3.2	-0.85	2300 m*	1900 m
S.E. Greenland, 66°15′–67° N Ammassalik Gribbon (1963)	940–1560 m	4.8	-0.95	1950 m*	
Present paper Table 2 and Figure 8 S.E. Greenland, 68°15′–69°N Kangersertuaq (Kangerlussuaq)	600–2400 m	2.1	-0.74	2250 m	2480 m
Present paper Table 7 and Figure 8 E. Greenland, 74°15′ N Schwarzenbach (1961)	610–2100 m	2.2	-0.71	2400 m	2100 m
	1200–2100 m**	3.1		2100 m	2100 m
	450–1400 m	3.3	-0.59	1800 m	1650 m

Table 10. Comparison of the rates of decrease of the number of species with altitude and of maximum altitudinal limits in different parts of Greenland. All calculations are by the present author based on selected sites. The altitudinal range is that between the lowest altitude of the lowest and of the highest sites.

* exceeds the highest mountains in the immediate area
** based only on sites above 1200 m
*** observed at 1900 m further inland

Evighedsfjord and comparable perhaps to the Kangersertuaq (Kangerlussuaq) area on the east coast. As with Gribbon's Ammassalik data, the figures for Evighedsfjord and Upernivik Ø do suggest a general lowering of altitudinal limits and a greater rate of decrease of the flora nearer the outer coast.

6.2   South Greenland

Gribbon led expeditions to the Kap Farvel area in 1971 and 1975 (Gribbon 1971, 1975). The first climbed on both sides of the middle part of Tasermiut fjords inland from Nanortalik (60° 15′ N). Six of his sites were above 700 m, his highest being at 1265 m. Using his data for all of these sites and a supplementary one referred to at the end of his botanical report, the calculated decrease in the number of species with altitude is 6.6 per 100 m and the altitudinal limit c. 1500 m. Gribbon, however, considered the limit to lie at about 1370 m. His 1975 expedition climbed chiefly in the area around Aappilattoq, to the south-east, and his montane sites ranged from 960–1430 m. Two were from summits: one at 1380 m yielding only one species, the other, at 1430 m, producing two. These figures certainly suggest a limit below 1500 m. Being nearer the outer coast the mountains are generally lower than those climbed in 1971 and, below 1200 m, they had an appreciably richer flora.

In 1972 an expedition from Leicester Polytechnic climbed mountains on the west side of Søndre Sermilik fjord towards its head and appreciably nearer the ice-cap than Gribbon's 1971 sites. Some plant collections were made and two species, *Ranunculus pygmaeus* and *Cardamine bellidifolia*, were noted at their highest site (A. Barbier 1973 in litt.) at 1700 m. The previous year, 1971, the Irish Greenland Expedition climbed around the edge of the ice-cap in the area between the head of Tasermiut and Lindenow Fjord and made a number of plant collections (J. Lynam 1971 in litt.). Two of these were from altitudes above 1500 m. Three species were noted at the lower site, 1650 m and appreciably above Gribbon's altitudinal limit, and about 12 were seen at the higher site, c. 1900 m on the south side of Point 2029 m midway between the two fjords. The discrepancy between these observations is considerable but there are a number of possible and not mutually exclusive explanations. It could be caused by the relatively low altitude of the mountains from which Gribbon made his collections: all lie below 1500 m. It would be interesting to know what the limits are on the higher peaks in this area, some of which ascend to over 1900 m. On the other hand it could be argued that the higher limit noted by the Irish expedition is perhaps due to a less oceanic climate around the heads of the fjords as well as the presence near the ice-cap of substantially more continuous areas of high land above 1900 m.

The rate of decrease of the flora with altitude derived from Gribbon's data is appreciably greater than that observed in any of the other areas and is no doubt a reflection of the extremely oceanic nature of this stormbound southern tip of Greenland.

The flora of two nunatak areas east of Frederikshåb (62° N and 63° N) has been described in a brief paper by Frederiksen (1971) using data collected by L. Bolt Clausen and herself and previously published data for the Jensen Nunatakker, particularly that of Gjærevoll (1968).[1] At the same time she briefly compared the data with Stocken's and Schwarzenbach's (1961) from east Greenland, Schwarzenbach's area lying in the northern part of the central fjord region between 74° and 75° N. Frederiksen's four sites cover a wide altitudinal range, from 700 to 1700 m, and lie from 3 to 23 km inland from the margin of the inland ice. Only two species of vascular plants were found at her highest site. In contrast 62 species were recorded from the Jensen Nunatakker which lie 25 km from the edge of the inland ice and range from 1200 to 1400 m. The spectrum of distribution types at the two areas is extremely similar, just over a third of the species being widespread arctic montane (1), with low-arctic (5) and low-arctic oceanic montane species (6), each contributing about 20%. The high proportion of Type 5 species is particularly significant and higher than that of any other area considered here.

6.3   East Greenland

Over a 50 year period, 1948–1998, F.H. Schwarzenbach made a detailed study of the altitude ranges of vascular plants in four areas of east Greenland, from Scoresby Land (71° 50′) to the nunatak zone west of Clavering Ø (74° 15′), and one in the extreme north-east in Kronprins Christian Land. His data appeared

---

1   The following species (Gjærevoll & Ryvarden 1977) should be added to Gjærevoll's list as published by Frederiksen: *Huperzia selaqo, Polygonum viviparum, Cerastium arcticum, Ranunculus pygmaeus, Draba lactea* (as *D. boecheri* Gjærevoll & Ryvarden), *D. norvegica, Tofieldia pusilla, Juncus trifidus, Luzula confusa, L. spicata, Carex bigelowii, C. capillaris, C. glacialis, C. norvegica, Festuca brachyphylla* (as *F. jenseni* Gjærevoll & Ryvarden), *Hierochloe ornantha* and *Trisetum spicatum*.

in two papers, the first (Schwarzenbach 1961) dealing with the nunatak zone, the second (Schwarzenbach 2000) with all five areas.

Using his data for the nunatak zone (Schwarzenbach 1961, pp. 55–57) and, as before, selecting only the richest sites but omitting the two valley ones, the calculated rate of decrease is 3.3 species per 100 m and the altitudinal limit 1900 m, substantialy higher than the highest observed record which is of *Poa abbreviata* at 1650 m (Gelting 1934, height amended by Schwarzenbach p. 20). The calculated height is certainly too high; it should be stressed, however, that Schwarzenbach had no sites above 1470 m and that the correlation coefficient, -0.59, is the lowest in Table 10. Schwarzenbach himself (p. 167) considered the limit to lie at about 1600 m and "considerably higher than on the outer coast". If this is a more realistic figure then, bearing in mind the latitude of the area, it is still very high, only 250 m lower than in the Kap Farvel area some 1600 km to the south.

It seems clear that, as in the Kangersertuaq and Ammassalik areas, the effects of latitude are offset by the continentality of the nunatak climate. This is strikingly demonstrated by the distribution-type spectrum of Schwarzenbach's nunatak sites which is quite unlike that from any other area with 80% of the species belonging to the widespread arctic (1), high-arctic (2), arctic continental (3) or medium-arctic montane (4) types.

6.4 Summary

Finally, in Table 11 are listed those species which were recorded from the highest site in each of the six areas represented in Table 10.

In his second paper Schwarzenbach (2000) discussed the maximum limit of vegetation and decided to record this not as the highest record of any species or that derived from the calculated rate of decrease but the highest altitude reached by any five species. His data indicate yet again the marked increase in altitudinal limits with increasing distance from the coast. In the Staunings Alper (72° N), for example, the east and north-eastern side facing the broad Kong Oscar Fjord shows an absolute limit of 1450 m whereas on the western side *Saxifraga cernua* reaches 2200 m and in Nathorst Land to the west *S. oppositifolia* and *Salix arctica* reach 2059 m, altitudes almost comparable to

SPECIES	1	2	3	4	5	6
Papaver radicatum	•	•	•	•	•	•
Saxifraga cespitosa	•	•	•	•	•	•
Cerastium arcticum/alpinum		•	•	•	•	•
Saxifraga cernua		•	•	•	•	•
S. oppositifolia	•		•	•		•
Poa glauca		•	•		•	•
Cardamine bellidifolia	•	•	•			•
Silene acaulis	•	•		•		
Luzula confusa		•				•
Saxifraga nivalis			•			•
Draba nivalis			•	•		
Minuartia rubella				•		•
Erigeron uniflorus				•	•	
Cystopteris fragilis					•	•
Luzula spicata		•				
Saxifraga foliolosa		•				
S. rivularis		•				
Oxyria digyna				•		
Antennaria canescens				•		
Arabis arenicola					•	
Festuca baffinensis					•	
Draba arctica					•	
D. subcapitata						•
Ranunculus glacialis						•
Campanula uniflora						•
Poa abbreviata						•
Melandrium apetalum						•
Poa arctica						•
Trisetum spicatum						•

Table 11. Species recorded from the highest site in each of the following areas.

1 S. Greenland: 64° 50′ N, Kap Farvel, J. Lynam (1971 *in litt.*), 1650–1900 m
2 W. Greenland: 64° 50′ N, Evighedsfjord, Gribbon (1965), 1370–1680 m
3 W. Greenland: 71° 15′ N, Upernivik Ø, Gribbon (1983 *in litt.*), 1850–1900 m
4 S..E. Greenland: 66° 15′–67° N, Ammassalik, Table 2, N13, 2200–2480 m
5 S.E. Greenland: 68° 15′–69° N, Kangersertuaq, Table 7, L1, 2500 m
6 E. Greenland: 74° 15′ N, Schwarzenbach (1961), 1350–1470 m

the limits given earlier in this paper for the mountains behind Ammassalik and Kangersertuaq, 750 and 370 km respectively further south. By contrast the highest altitude recorded by Schwarzenbach in Kronprins Christian Land, 950 km to the north, was 955 m.

The commonest species at the higher altitudes are, not surprisingly, very similar in the two areas discussed in this paper and those listed by Schwarzenbach (2000, p. 149) where his 42 commonest species are listed in order of decreasing altitudinal limits. Of the 11 commonest species in the middle and upper zones in the Ammassalik area only one, *Oxyria digyna*, is absent from Schwarzenbach's first 18 species, all of which have limits at or above 1200 m, while of the 14 species in the upper zone in the Kangersertuaq area only two, *Ranunculus glacialis* and *Draba norvegica*, are absent. The numerous species reported at surprisingly high altitudes in the present paper must surely be serious contenders in any discussion of survival in Greenland during the last glaciation (Halliday 2002).

Concluding remarks

The flora decreases steadily with altitude at the rate of about two species per 100 m. The highest observed limits were 2480 m in the southern area and 2100 m in the northern. These two areas include the two highest mountains in Greenland, and indeed in the Arctic as a whole. It is therefore very likely that many of the species occurring over about 1500 m are here at their highest recorded altitude in the Arctic.

With increasing altitude the spectrum of biological distribution types becomes progressively more limited and in the upper zone, above 1800 m, nearly two thirds of the flora belong to the widespread arctic montane type.

The continentality of these inland areas allows six species of the mid- and high-arctic to occur at high altitudes and much further south than was previously known.

Acknowledgements

I am greatly indebted to the many leaders of the expeditions featured in this paper for their willingness to co-operate with the botanical collecting, and also to those who actually spared time to engage in what was for many of them a novel experience. Thanks are also due to Dr C.K. Brooks, Dr B. Fredskild, Dr E. Hoch and Dr E. Steen-Hansen for comments and information, and to Ruth Berry for technical assistance with the figures.

APPENDICES

Appendix I: Lowland and coastal botanical collecting sites of expeditions featured in Table 1

YEAR	NAME	SITES	COMMENTS
1935–36	British East Greenland Expedition	Ammassalik, Leif Ø; Tuttilik; outer and mid-Kangersertuaq (Kangerlussuaq); Mikip Kangertiva (Miki Fjord); 2-miippugut (J.C. Jacobsen Fjord); 5-imiippugut (Nansen Fjord)	*Potentilla tridentata*★, *Thymus praecox*★ *Pseudorchis albida*★
1963	Scottish East Greenland Expedition	Ammassalik, Kulusuk; head of Tasiilaq (Tasîssârssik)	Gribbon (1968a)
1966	Royal Naval East Greenland Expedition	Kulusuk; head of Tasiilaq	
1966	Schweizer Grönland-Expedition	head of Kangersertivattiaq (Kangertítivatsiaq)	Elsley & Halliday (1971)
1967	Imperial College East Greenland Expedition	head of Kangersertivattiaq	Elsley & Halliday (1971)
1969	London University Graduate Mountaineering Club East Greenland Expedition	head of Kangersertivattiaq	most of the collection abandoned, *Equisetum arvense* new to area
1971	Expedition to the Steenstrup area of East Greenland	head of Kangersertivattiaq; north side of Kialiip Tasiilaa (Tasîlaq)	8 additional species 17 species recorded
1971	Watkins Mountains Expedition Phase II	Kap Garde, Kivioq Fjord, Kap Normann	
1972	Westminster East Greenland Expedition	Ammassalik; Nordre Aputiteeq; outer Kangersertuaq	*Woodsia alpina*, *Asplenium viride*★, *Carex capitata*, *C. misandra*, *C. rufina*
1974	Westminster East Greenland Expedition	north side of Kialiip Tasiilaa	47 additional species
1979	Durham University Polar East Greenland Expedition	Ammassalik north side of Kialiip Tasiilaa	41 additional species including *Asplenium viride*, *Botrychium boreale*★, *Gnaphalium norvegicum*★, *Hieracium groenlandicum*★, *Pseudorchis albida*

★   New northern limit in east Greenland

Appendix II: Vascular plants featured in Tables 2 (Ammassalik area) and 3 (Kangersertuaq area), with maximum altitudes and distributional★ and taxonomic† comments

Agrostis mertensii 2 1060 m (G17)

Alchemilla glomerulans 2 600 m (I5)

Antennaria canescens 2 2200 m (N13) 3

    porsildii 2 900 m (N8)

Arabis alpina 2 1550 m (W9) 3

    ★*arenicola* 3 2000 m (L1)

Arnica angustifolia 2 940 m (G2) 3

Bartsia alpina 2 1060 m (G17)

Campanula gieseckiana 2 1560 m (G8) 3

    uniflora 2 1780 m (S13) 3

Cardamine bellidifolia 2 2000 m (D10) 3

Carex bigelowii 2 1460 m (D11) 3

    capillaris 3 850 m (A14)

    capitata 2 1060 m (G17)

    glacialis 2 1400 m (G6) 3

    lachenalii 2, 3 1300 m (B1)

    macloviana 2 760 m (D3)

    maritima 3 1740 m (A4)

    misandra 3 1300 m (B1)

    nardina 2 2100 m (W7) 3

    norvegica 2 1510 m (G7) 3

    scirpoidea 2, 3 1300 m (B1)

Cassiope tetragona 2 1200 m (D16) 3

†*Cerastium arcticum* 2 2200 m (N13) 3

    cerastoides 2, 3 1200 m (W1)

Chamerion latifolium 2 1900 m (A2) 3

Cystopteris fragilis 2, 3 2000 m (L1)

Deschampsia alpina 2 940 m (G2)

Diapensia lapponica 2 1050 m (T1)

Diphasiastrum alpinum 2 960 m (G3)

★*Draba arctica* 3 2000 m (L1)

    aurea 2 1000 m (G4) 3

    nivalis 2 2200 m (N13) 3

    norvegica 2 2000 m (D10) 3

    ★*subcapitata* 2 1600 m (M1) 3

Dryas integrifolia 2 1060 m (G17)

Empetrum nigrum

    ssp. *hermaphroditum* 2 1350 m (G5) 3

Epilobium sp. 2 600 m (I5)

Equisetum arvense 3 700 m (K1)

    variegatum 2, 3 700 m (K1)

★*Erigeron compositus* 2 1300 m (N11) 3

    humilis 2 1780 m (S13) 3

    uniflorus 2 2200 m (N13) 3

Eriophorum scheuchzeri 2 810 m (G16)

Euphrasia frigida 2 1500 m (W12) 3

★*Festuca baffinensis* 3 2000 m (L1)

    brachyphylla 2 2000 m (D10) 3

    vivipara 2 1450 m (W11)

Gentiana nivalis 2 1000 m (G4) 3

Gnaphalium norvegicum 2 600 m (I5)

    supinum 2 1050 m (T1)

Harrimanella hypnoides 2 1350 m (G5) 3

Hieracium alpinum 2 1200 m (D16)

Hierochloe alpina 3 1450 m (W5)

Huperzia selago 2, 3 1300 m (B1)

Juncus trifidus 2 1350 m (G5) 3

Juniperus communis 2 1000 m (G4)

Loiseleuria procumbens 2 1000 m (D16)

Luzula confusa 2 2000 m (D10) 3

  †multiflora 2 600 m (I5)

  spicata 2 2100 m (W7) 3

Lycopodium annotinum 2 960 m (G5)

*†Melandrium affine 2, 3 1760 m (W15)

Minuartia biflora 2 2000 m (D10) 3

  rubella 2 2200 m (N13) 3

  *stricta 2 2000 m (S14) 3

Oxyria digyna 2 2200 m (N13) 3

*Papaver radicatum 2 2300 m (L3) 3

Pedicularis flammea 2 880 m (S3)

  hirsuta 2 1000 m (G4)

Phippsia algida 2 1800 m (N12) 3

Phleum commutatum 2 600 m (I5)

Phyllodoce caerulea 2, 3 1200 m (W37)

*Poa abbreviata 3 1830 m (W13)

alpina 2 1500 m (D4) 3

arctica 2 2000 m (L2) 3

glauca 2 2000 m (L1) 3

Polygonum viviparum 2 1680 m (D6) 3

Potentilla crantzii 2 1700 m (N9) 3

  hyparctica 3 1450 m (W5)

  *†nivea / hookeriana 2 1900 m (A2) 3

  tridentata 2 960 m (G3)

Ranunculus glacialis 2, 3 2100 m (B2)

  pygmaeus 2 1750 m (N6) 3

Rhodiola rosea 2 1900 m (N14) 3

Rhododendron lapponicum 2 1350 m (G5) 3

Rumex acetosella 2 1050 m (T1)

Sagina intermedia 3 1000 m (B3)

Salix glauca 2 1500 m (D4) 3

  herbacea 2 1600 m (N5) 3

Saxifraga cernua 2 2200 m (N13) 3

  cespitosa 2 2200 m (N13) 3

  hyperborea 2 1550 m (W9)

  nivalis 2 2000 m (D10) 3

  oppositifolia 2 2200 m (N13) 3

  paniculata 2, 3 1300 m (B1)

  rivularis 3 1300 m (W3)

  tenuis 2 1750 m (D9) 3

Sedum annuum 3 700 m (K1)

  villosum 3 700 m (K1)

Sibbaldia procumbens 2 1510 m (G17) 3

Silene acaulis 2 2200 m (N13) 3

†Stellaria longipes sensu lato 2, 3 700 m (K1)

Taraxacum sp. 2 1900 m (N14) 3

Thymus praecox 2 1060 m (G17)

Tofieldia pusilla 2 1060 m (G17) 3

Trisetum spicatum 2, 3 1950 m (A3)

Vaccinium uliginosum 2 1400 m (G6) 3

Veronica alpina 2 1000 m (G4) 3

  fruticans 2, 3 1190 m (A8)

Viscaria alpina 2, 3 1190 m (A8)

Woodsia alpina 2 1800 m (N12) 3

  glabella 2 1500 m (W5)

  ilvensis 2 1200 m (D16) 3

Distributional comments

The following species previously unknown south of Scoresby Sund (70° N) now have new southern limits in East Greenland:

Minuartia stricta loc. N14, 66° 38′ N (Figure 7a)
Melandrium affine loc. S9, 66° 39′ N (Figure 7b)
Draba subcapitata loc. M1, 66° 55′ N (Figure 7c)
Draba arctica loc. W1, 68° 31′ N (Figure 7f)
Poa abbreviata loc. A4, 68° 49′ N (Figure 7e)
Festuca baffinensis loc. A4, 68° 49′ N (Figure 7d)

The occurrence of *Minuartia stricta* is particularly remarkable as this is essentially a lowland species in central east Greenland, the maximum altitude given by Schwarzenbach (2000) being 510 m.

New northern limits for species from lowland sites are given in Appendix I.

The records for the following species add substantially to their previously very disjunct distributions south of Scoresby Sund:

Arabis arenicola (Figure 6b)
Papaver radicatum (Figure 6a)
Potentilla nivea/hookeriana (Figure 6c)
Erigeron compositus (Figure 6d)

The records of *Arabis arenicola* (Figure 6b) are most remarkable. This is generally considered to be a lowland species of stream- and riversides. Böcher et al. (1978) give its distribution in east Greenland as 70°–72° N. Since then its northern limit was extended in 1980 by the present author to the south side of Lindeman Fjord (74° 38′ N). To the south it was found at three sites in the Kangersertuaq area and later by Bay (1993) in south-east Greenland near sea-level inland of Skjoldungen (63° 28′ N).

Taxonomic comments

Abbreviations in bold type are international herbarium abbreviations.

Cerastium arcticum
Although some of the literature records (e.g. Gribbon 1968a) are of *C. alpinum*, all the material in **E** is referable to *C. arcticum* subsp. *hyperboreum*. All the nunatak records are therefore included in this paper under *C. arcticum*. Many of the specimens were originally identified as *C. arcticum* var. *vestitum* Hult., a taxon which is now included in subsp. *hyperboreum*. The plants are rather densely tufted, the leaves are very hairy but lack the conspicuous tuft of white hairs so characteristic of *C. alpinum*, and the large flowers are mostly solitary or paired.

Chamerion latifolium (L.) Holub
The genus *Chamerion* (Raf.) Raf. replaces the familiar but illegitimate name *Chamaenerion* L.

Erigeron uniflorus sensu lato
All the material in **E** is referable, as one might expect, to *E. eriocephalus*, having robust stems and large, wide, densely hairy capitula, with usually spreading involucral bracts.

Luzula multiflora
The material from the east side of Kattilersorpia ('Glacier de France', loc. G15) is of subsp. *frigida*.

Melandrium affine
The statement by Böcher et al. (1978) that *M. triflorum* occurs near Mont Forel is based on the specimen in **C** collected by B.K. Porter (loc. A1). B. Fredskild agrees with the present author that the material is inadequate for a definite identification, but a duplicate in **E** has immature seeds which are nevertheless distinctly winged and so referable to *M. affine*, the commoner of the two species in east Greenland.

Of the other two nunatak collectons in **E**, one, from the same site as the above but collected the following year, is vegetative. The other is from Kangersertuaq and has well-developed winged seeds. It was collected by

P. Wager, almost certainly from the west side of the fjord. There is therefore no evidence that *M. triflorum* occurs south of Scoresby Sund.

Melandrium apetalum

This species is surprisingly included by Lodge (1974) from both the coastal site at Kialiip Tasiilaa (Tasîlaq) and his nunatak site W5. It was not seen by the 1979 Durham University Expedition which studied the area in considerable detail, nearly doubling the number of species previously recorded. The whereabouts of the Westminster Expedition material is unfortunately unknown but it seems highly likely that both records refer to *M. affine* and they have been mapped as such in Figure 7b.

However, the occurrence of *M. apetalum* cannot be entirely dismissed since the southernmost authenticated records on the east coast are from 820 m above Torvgletscher, to the south of Scoresby Sund, and near sea-level at Kap Dalton (Halliday et al. 1974).

Potentilla nivea/hookeriana

An examination of herbarium material suggests that most of the material is referable to *P. nivea sensu stricto*. This is the only segregate represented at the isolated localities in the Skjoldungen area (c. 63° 30′ N) and according to Feilberg (1984) it is the only one in southernmost Greenland. However, plants of typical *P. hookeriana* subsp. *chamissonis* and intermediates, with the long straight petiolar hairs of the former but with an underlying felt of fine, crumpled *P. nivea* hairs, occur in both the Ammassalik and Kangerlussuaq areas. The localities from which herbarium specimens have been examined are listed below.

P. nivea sensu stricto

Ammassalik area: G3 (**STA**), D2 & S4 (**C**), N7 (**E**), N8 (**C**), also inland from Qíngertivag (**C, E**) and Qáqarssuaq (**C**)
Kangersertuaq area: W5 (**BM**), A5 (**E**), A7 (**E**), also Mudderbugt (**C, E**), Watkins Fjord (**C, E**) and Mikis Fjord (**C**)

P. hookeriana subsp. chamissonis

Ammassalik area: G3 (**STA**), I1 (**E**)
Kangersertuaq area: A5 (**E**)

Intermediate between *P. hookeriana* subsp. *chamissonis* and *P. nivea*

Ammassalik area: D6 (**C**), A2 (**E**), T1 (**E**)

Kangersertuaq (Kangerlussuaq) area A7 (**E**)

Stellaria longipes sensu lato

The material from the east side of Kattilersorpia (loc. I5) was clearly referable to *S. edwardsii* (Halliday et al. 1974), here at its southern limit in east Greenland.

Appendix III: Lists of bryophytes and lichens

Bryophytes and lichen samples were collected by the expeditions listed below, each prefaced by a different letter for ease of reference. Further information on the expeditions and sites is given in the main text Table 1 and in sections 4.1 and 4.2, except for those sites for which there are no vascular plant records. For these the localities, co-ordinates and collectors are given here.

Expeditions and collection sites

In the main text, both the 1935–36 British East Greenland Expedition to the northern area and the 1974 Westminster East Greenland Expedition to the southern area are designated with the letter W; and both the 1969 London University Graduate Mountaineering Club East Greenland Expedition to the southern area and the 1988 British East Greenland Expedition to the northern area are designated with the letter L. Since the northern and southern sites are treated together in this section, the expeditions to the northern areas are here indicated by the addition of a superscript ^{N} after their designators in order to avoid confusion between them. The 1935–36 expedition is thus W^N, and the 1988 expedition is L^N. The 1978 Westminster East Greenland Expedition is designated W^F to distinguish it from the 1974 Westminster East Greenland Expedition.

$\mathbf{W}^N$ British East Greenland Expedition 1935–36

The 11 localities referred to by Mackenzie Lamb (1940) lack both collectors' names, coordinates and dates. Sites 41–46 are not in the main text.

W^N38   East side of Christian IV Gletsjer: Watkins Bjerge, 1700 m, 1800 m
W^N40   East side of Christian IV Gletsjer: Watkins Bjerge, scree ridge, 3048 m
= Lichenbjerge, 68° 56′ W, 30° 00′ W, L.R & H.G. Wager, 1935
W^N3   Near head of Sorgenfri Gletsjer, 1200 m

W^N3 West side of Christian IV Gletsjer: "Windy Gap", 1300 m

W^N5 West side of Christian IV Gletsjer: "Coal corner", 1300 m

W^N7 (= W^N10) West side of Christian IV Gletsjer: Lindbergh Fjelde: 1400 m, 1450 m

W^N11 West side of Christian IV Gletsjer: Lindbergh Fjelde, 1500 m

W^N41 West side of Christian IV Gletsjer: "Snow Dome", 2225 m, A. Courtauld & E.C. Fountaine, 1935 = Snøkolpen, 68° 47′ N, 31° 08′ W

W^N42 Seward Plateau: "Observation Nunatak" at N end of Prinsen av Wales Bjærge, 1800 m, coordinates not known, W.A. Deer & E.C. Fountaine, 1936

W^N43 "Nunatak Camp", 25 km north-west of Prince of Wales range, 1400 m, coordinates not known, W.A. Deer & E.C. Fountaine, 1936

W^N44 Near head of Kangersertuaq Gletscher, "Triangular Nunataks", 1300 m, W.A Deer & E.C. Fountaine, 1936 = Trekantnunatakker, 68° 43′ N, 33° 55′ W

W^N45 West side of Kangersertuaq: "Panorama Nunatak", 1200 m , W.A. Deer & E.C. Fountaine, 1936 = Panoramanunatakker, 68° 13′ N, 34° 28′ W

W^N46 (=R1) West side of Kangersertuaq: "Ice fall Nunatak", near head of Nordre Parallel Gletscher, 1150 m, 67° 53′ N, 33° 15′ W, W.A. Deer & E.C. Fountaine, 1936

D   Schweizerisch-Deutsche-Grönland-Expedition 1963

N   Royal Naval East Greenland Expedition 1966

S   Schweizer Grönland-Expedition 1966

I   Imperial College East Greenland Expedition 1967

A   Army East Greenland Expedition 1968

L   London University Graduate Mountaineering Club East Greenland Expedition 1969

T   Expedition to the Steenstrup area of East Greenland 1971

W   Westminster East Greenland Expedition 1974

W^F   Westminster East Greenland Expedition 1978
(Kronprins Frederik Bjerge, southern part)

W^{F}1   66° 46′ N, 34° 37′ W, 905 m
W^{F}2   66° 55′ N, 34° 32′ W, 1215 m
W^{F}3   66° 48′ N, 34° 25′ W, 1200 m
W^{F}4   67° 26′ N, 34° 17′ W, 1560 m
W^{F}5   67° 46′ N, 34° 15′ W, 1665 m
W^{F}6   67° 51′ N, 34° 17′ W, 2050 m
W^{F}7   67° 47′ N, 34° 28′ W, 1915 m
W^{F}8   67° 40′ N, 34° 42′ W, 2230 m
W^{F}9   67° 25′ N, 34° 47′ W, 2240 m
W^{F}10   66° 58′ N, 34° 47′ W, 1720 m
W^{F}11   67° 09′ N, 34° 42′ W, 2260 m
W^{F}12   66° 42′ N, 34° 43′W, 1030 m
W^{F}13   67° 53′ N, 34° 25′ W, 1850–1900 m
W^{F}14   66° 55′ N, 34° 52′ W, 1705 m

M   Durham University Polar East Greenland Expedition 1979

L$^{(N)}$ British East Greenland Expedition 1988

R   Northern Group Greenland Expedition 1990
(Kronsprins Frederik Bjerge, northern part)

R1   Panoramanunataker, 68° 13′ N, 34° 28′ W, 2300 m
R2   Nunatak 30 km south-east of Panoramanunataker, 68° 04′ N, 33° 57′ W, 2300 m
R3   Nunatak 10 km south-west of R2, 68° 00′ N, 34° 05′ W, 2150 m

C   British Mountaineering Expedition to East Greenland 1992

Collected samples of bryophytes and lichens

The following three herbaria are designated by conventional abbreviations in **bold** in this section: All expedition records are not necessarily represented by herbarium material.

BM	Natural History Museum, London
C	Botanical Museum, University of Copenhagen
E	Royal Botanic Garden, Edinburgh

Bryophytes

Nomenclature follows Hill et al. (2006). The list is based on material collected by the twelve expeditions listed below. The maximum altitude is given for each species. Material collected on the British East Greenland Expedition (1935–1936, **BM**) has not been identified. Bryophytes collected on the 1972 Westminster East Greenland Expedition were passed to the late W.D. Foster for identification. Neither identifications nor the whereabouts of the material are known.

W[N]	material not identified, **BM**
D	det. K. Holmen, **C**?
N	det. G. Halliday, **E**
S	det. G. Halliday, **C**?
I	det. G. Halliday, **E**
A	det. G. Halliday, **E**
L	det. G. Halliday, **E**
T	det. G. Halliday, **E**
W	det. E. Lodge, location unknown
M	det. G. Halliday, **E**
L[N]	det. G. Halliday, **E**
R	leg. J. Richardson, det. G. Halliday, **E**

Mosses

Aulacomnium palustre I5 940 m; G2 (Figure 3)
    M. Lenarčič, 1978

Bartramia ithyphylla I1; W2, 5, 7 2100 m, 12, 13

Blindia acuta I5; W2, 5 1500 m, 12

Bryum sp. I1; L1 1700 m

Dicranoweisia crispula N5, 11, 13 2200 m; L-
    west side of K.I.V. Steenstrup Nordre
    Bræ, 66° 47′ N, 35° 25′ W, 2000 m; W2,
    5, 12, 13

Dicranum fuscescens I5; N4; W5, 7 2100 m,
    10, 12

Distichium capillaceum S5, 6 1100 m

Ditrichum flexicaule I2 1700 m

Drepanocladus polygamus I4 2050 m

Grimmia doniana A2; R1, 2 2300 m

G. longirostris det. H. Blom N11 1300 m

Hypnum revolutum D9; I1, 3; L$^{(N)}$1 2100 m

Isopterygiopsis pulchella I1, 2 1700 m

Onchophorus virens I1 1470 m

Orthotrichum speciosum I2, L$^{(N)}$1 2000 m, det.
    R. Ochyra

Philonotis tomentella I2 1700 m, 5

Pogonatum dentatum D1; W3, 12 1500 m

P. urnigerum S12, 15 2000 m; I3, 5

Pohlia cruda I1; W5, 7, 10; L$^{(N)}$1 2100 m

P. nutans I1 1470 m

P. obtusifolia S3 880 m

Polytrichastrum alpinum I1, 4; L1, 2; N5; W2,
    5, 7 2100 m, 8, 10, 12, 13; M1

Polytrichum strictum S3; I5; W3 1250 m

P. piliferum N5; S1, 2, 9; I1, 3; T1; W1, 2, 3, 5, 7
    2100 m, 10, 12, 13

Racomitrium canescens N5; S4, 8, 12, 15; I1, 2,
    3, 4; A2; T1; W1, 2, 3, 5, 7 2100 m, 8, 9,
    10, 11, 12, 13; M1

R. lanuginosum I1, 3 1800 m; N5, 8

Sanionia uncinata S1; I5 600 m

Schistidium apocarpum s.l. W1, 2, (4)
    66° 42′ N, 34° 40′ W, 900–950 m, 5, (6)
    66° 59′ N, 35° 28′ W, 1900–1950 m, 7, 10

S. frigidum det. H. Blom I1, 4 2050 m

S. venetum det. H. Blom I2 1700 m

Scorpidium revolvens I5 600 m

Sphagnum capillaceum I5 600 m

Tortella fragilis I2 1700 m

Tortula hoppeana I3 1800 m

T. ruralis I3 1800 m

Warnstorfia sarmentosum I5 600 m

(*Meesia triquetra* W2, 3, 5, 7, 10: the presence
    of this species at these altitudes seems
    so unlikely that the records are prob-
    ably the result of misidentification)

Liverworts

Anthelia julacea W2, 3, 5 1500 m, 12, 13

Cephaloziella sp. I1, 3 1800 m

Lichens

Nomenclature mostly follows Santersson et al. (2004). The list is based on material collected by the expeditions listed below. The most important was undoubtedly the 1935–1936 British East Greenland expedition to the Kangersertuaq region (W^N), the specimens of which were identified and the records published by Lamb (1940). The maximum altitude is given for each species.

W^N det. I. Mackenzie Lamb, **BM**

D   det. K. Holmen, **C?**

N   det. D.L. Hawksworth & K.A. Kershaw, **C**, **herb. D.L. H.**

S   det. E. Steen Hansen, **C?**

L   det. E. Steen Hansen, **C**

W^F det. P. James, **BM**

R   leg. J. Richardson, det. E. Steen Hansen, **C**

Alectoria nigricans N8 900 m

Arthrorhaphis citrinella N13 2200 m; W^F5

Aspicilia disserpens W^N3, 5 1450 m

Bryoria chalybeiformis N6 1750 m, 7, 8

Candelaria cf concolor N8, 11, 13 2200 m

Candellariella kuusamoensis W^F12 1030 m

C. vitellina W^N7 1400 m

Cetraria. islandica N4, 8; S1, 4 (=D2) 1020 m

Cladonia cf borealis N8, 13 2200 m

C. chlorophaea N11 1300 m

Dermatocarpum miniatum W^N11 1500 m, 38

Flavocetraria cucullata N8; S6 1100 m

F. nivalis N4, 6 1750 m, 8, 11; S1, 4(=D2), 9, 12; W^F12

Lecanora atrosulphurea W^N40 1300 m

L. intricata W^F7 1915 m

L. polytropa W^N11, 44; W^F1, 5, 7 1915 m

Lecidiella anomaloides W^N3, 7 1400 m, 40, 44

L. sp. W^N5, 9 2240 m

Lepraria neglecta N8 900 m

L. vouaunii W^F10, 1720 m

L. sp. W^N38, 1250 m

Megospora verrucosa W^N7 1400 m

Miriquidica garovaglii W^N3 1300 m

Peltigera didactyla N12 1800 m

Physconia muscigena W^N7 1400 m

Pleopsidium chlorophanum W^N7, 10, 11, 38, 40, 44; N12; W^F2, 3, 6, 9, 12; R1 2300 m

Porpidia melinodes W^N15 1760 m; W^F12

Pseudephebe minuscula W^N40, 44, 46; N7; W^F4, 8 2230 m; R3

P. pubescens W^N7; N8; L3; W^F1, 8, 12 2300 m

Rhizocarpon copelandii W^F9 2240 m

R. flavum W^N3; W^F1, 7 1915 m

R. frigidum W^F3 1200 m

R. genuatum W^N7 1400 m

R. inarense W^F4, 9 2240 m, 10, 12

R. sp. W^F7, 8 2230 m

R. superficiale W^N3; W^F4 1560 m, 12

Rhizoplaca chrysoleuca W^N3, 7, 11 1500 m, 44, 45

R. melanophthalma W^N5, 11, 38, 40, 41 2225 m, 42, 45; W^F1

Solorina bispora W^N38 1250 m

S. crocea N6; S4(=D2), 9; W^F12; C1 1750 m

Sporastatia testudinea W^F8 2230 m

Staurothele areolata W^N38 1250 m

S. fuscocuprea W^N5 1450 m

Stereocaulon alpinum N5 1600 m, 11

S. botryosum W^N43, 44; N8; W^F1, 5, 6 2500 m

S. condensatum W^F12 1030 m

S. rivulorum W^N3, 7 1400 m, 46

Thamnolia vermicularis N8, 13 2200 m

Toninia cf havaasii W^F9 2240 m

Tremolecia atrata W^N3 1300 m

Umbilicaria arctica W^F2, 12 1215 m

U. cylindrica N7, 13; W^F4, 9 2240 m

U. decussata W^F7, 9 2240 m, 12

U. hyperborea N7 1100 m, 8

U. krascheninnikovii W^N3 1300 m

U. cf lyngei W^N44 1300 m

U. cf rigida N13 2200 m

U. sp. W^F8, 11; R1 2300 m

U. vellea N7 1100 m

U. virginis W^N5, 7, 11, 38, 43, 44; N4; W^F4; R3 2150 m

Usnea sphacelata W^N7; N8; S3; L3; R1 2300 m

Verrucaria polycocca W^N3 1300 m

Xanthoria elegans W^N7, 11, 38, 40, 42, 43, 44; N7, 8, 11, 13; W^F2, 9 2240 m, 10, 15

References

Unless cited in the text the expedition reports referred to in Table 1 are excluded from the following list. Expedition reports seldom give the date of publication and this has therefore been assumed to be, probably wrongly in most cases, the year of the expedition.

Bay, C. 1992. A phytogeographical study of the vascular plants of northern Greenland – north of 74 northern latitude. – *Meddr. Grønland, Biosci. 36:* 102 pp.

— 1993. Grønlands Botaniske Undersøgelse 1992: 26–37. – Botanisk Museum, Copenhagen.

Bruun, H.H. et al. 2006. Effects of altitude and topography on species richness of vascular plants, bryophytes and lichens in alpine communities. – *J. of Vegetation Science* 17: 37–46.

Böcher, T.W. 1933a. Phytogeographical studies of the Greenland flora based upon investigations of the coast between Scoresby Sound and Angmagssalik. – *Meddr. Grønland* 104(3): 56 pp.

— 1933b. Studies on the vegetation of the east coast of Greenland between Scoresby Sound and Angmagssalik (Christian IX's Land). – *Meddr. Grønland* 104(4): 134 pp.

— 1938. Biological distributional types in the flora of Greenland. – *Meddr. Grønland* 106(2): 339 pp.

— 1954. Oceanic and continental vegetational complexes in South-west Greenland. – *Meddr. Grønland* 148(1): 336 pp.

— 1959. Floristic and ecological studies in Middle West Greenland. – *Meddr. Grønland* 156(5): 68 pp.

— 1963. Phytogeography of middle West Greenland. – *Meddr. Grønland* 148(3): 289 pp.

— et al. 1978. Grønlands flora. – P. Haase & Sønner, Copenhagen: 326 pp.

Courtauld, A. 1936. A journey in Rasmussen Land. – *Geogrl. J.* 88: 193–215.

Daniëls, F.J.A. & Molenaar, J.G. de 1970. Rare plants from the Angmagssalik District, Southeast Greenland. – *Bot. Tidsskr.* 65: 252–253.

Devold, J. & Scholander, P.F. 1933. Flowering plants and ferns of Southeast Greenland. – *Skr. Svalbard Ishavet* 56: 209 pp.

Elsley, J.E. & Halliday, G. 1971. Some plant records from Southeast Greenland. – *Meddr. Grønland* 178(8): 15 pp.

Feilberg, J. 1984. A phytogeographical study of South Greenland. Vascular plants. – *Meddr. Grønland, Biosci.* 15: 72 pp.

Frederiksen, S. 1971. The flora of some nunataks in Frederikshåb district, West Greenland. – *Bot. Tidsskr.* 69: 60–68.

Fredskild, B., Holmen, K. & Jakobsen, K. 1978. *Grønlands Flora* (3rd edition). – Haase & Søns Forlag, Copenhagen: 327 pp.

Gelting, P. 1934. Studies on the vascular plants of East Greenland between Franz Joseph Fjord and Dove Bay. (Lat. 73° 15′–76° 20′ N.). – *Meddr. Grønland* 101(2): 340 pp.

Gjærevoll, O. 1968. Det blomster på nunatakkene. – *Polarboken* 1968: 9–39.

— & Ryvarden, L. 1977. Botanical investigation on J.A.D. Jensens Nunatakker in Greenland. – *K. norske Vidensk. Selsk. Forh. Skr.* 4: 40 pp.

Gribbon, P.W.F. 1968a. Altitudinal zonation in East Greenland. – *Bot. Tidsskr.* 63: 342–357.

— 1968b. Distributional-type spectra in Greenland montane localities – *Bot. Notiser* 121: 501–504.

— et al. 1963. The Scottish East Greenland Expedition 1963. General Report: 52 pp.

— 1965. The University of St Andrews West Greenland Expedition 1965. General Report: 60 pp.

— 1967. The University of St Andrews West Greenland Expedition 1967. General Report: 64 pp.

— 1969. The University of St Andrews West Greenland Expedition 1969. General Report: 66 pp.

— 1971. The University of St Andrews South Greenland Expedition 1971. General Report: 82 pp.

— 1975. The University of St Andrews South Greenland Expedition 1975. General Report: 98 pp.

Halliday, G. 1967. In Royal Naval East Greenland Expedition 1966 Report. – Plymouth: 76 pp.

— 2002. The British Flora in the Arctic. *Watsonia* 24: 133–144.

—, Kliim-Nielsen, L. & Smart, I.H.M. 1974. Studies on the flora of the North Blosseville Kyst and on the hot springs of Greenland. – *Meddr. Grønland* 199(2): 49 pp.

— et al. 1981. British North-east Greenland Expedition 1980. Report. – University of Lancaster: 52 pp.

Hawksworth, D.L., James, P.W. & Coppins, B.J. 1980. Checklist of lichen-forming lichenicolous and allied fungi. – *Lichenologist* 12(1): 1–115.

Hill, M.O. et al. 2006. An annotated checklist of the mosses of Europe and Macaronesia. – *J. of Bryology* 28(3): 198–267.

Hoch, E. 1991. Palaeontological data on life in the intra-cratonic Kangerdlugssuaq basin. In C.K. Brooks & T. Stærmose (eds). *Kangerdlugssuaq Studies. Processes at a Rifted Continental Margin* II. Geologisk

Centralinstitut, University of Copenhagen, 45–52.

— 1992a. First Greenland record of the shark genus *Ptychodus* and the biogeographic significance of its fossil assemblage. – *Palaeogeography, Palaeoclimatology, Palaeoecology* 92: 277–281.

— 1992b. Palaeontology of the North Atlantic region: data and inferences on life in the Kangerdlugssuaq area before the continent severed. In C.K. Brooks, E. Hoch & A.K. Brantsen (eds). *Kangerdlugssuaq Studies. Processes at a Rifted Continental Margin* III. Geologisk Museum, Univeresity of Copenhagen, pp. 24, 104–110, Plate 1.

Kruuse, Chr. 1912. Rejser og botaniske Undersøgelser i Østgrønland mellem 65° 30′ og 67° 20′. – *Meddr. Grønland* 49: 307 pp.

Lamb, I.M. 1940. Lichens from East Greenland, collected by the Wager expedition 1935–36. – *Nyt Mag. Naturvid.* 80: 263–283.

Lodge, E. 1974. In Westminster East Greenland Expedition 1974 Report: 30 pp.

Manley, G. 1938. Meteorological observations of the British East Greenland Expedition, 1935–36, at Kangerdlugssuak, 68° 10′ N, 31° 44′ W. – *Q. Jl. R. met. Soc.* 64: 253–276.

Matthews, D.W. 1979. A buried Tertiary pluton in East Greenland? – *Bull. Geol. Soc. Denmark* 28: 17–20.

Santersson, R., R. Moberg, A. Nordin, T. Tonsberg & O. Vitikainen. 2004. *Lichenforming and lichenicolous fungi of Fennoscandia*. – Uppsala University: 359 pp.

Schwarzenbach, F.H. 1961. Botanische Beobachtungen in der Nunatakkerzone Østgrønlands zwischen 74° und 75° N.Br. – *Meddr. Grønland* 163(5): 172 pp.

— 2000. Altitude distribution of vascular plants in mountains of East and Northeast Greenland. – *Meddr. Grønland, Biosci.* 50: 193 pp.

Seward, A.C. & Edwards, W.N. 1941. Fossil plants from East Greenland. *Anns. Mag. Nat. Hist. Ser.* 11: 8, 169–176.

Wager, H.G. 1938. Growth and survival of plants in the Arctic. – *J. Ecol.* 26: 390–410.

Wager, L.R. 1937. The Kangerdlugssuak region of East Greenland. – *Geogrl. J.* 90: 393–425.

— 1947. Geological investigations in East Greenland. Part IV. The stratigraphy and tectonics of Knud Rasmussens Land and the Kangerdlugssuaq Region. – *Meddr. Grønland* 134(5): 62 pp.

Woolley, S. 2004. Greenland ventures. – Athena Press, London: 226 pp.

Young, R. 1980. In Jenkins, W.G. Narssarssuaq, South Greenland. 1972 and 1976 Expedition Results. Brathay Exploration Group, Field Studies Report No. 9. – Brathay Hall Trust, Ambleside: 117 pp.

Monographs on Greenland | *Meddelelser om Grønland*

Manuscripts should be sent to:

Museum Tusculanum Press
Dantes Plads 1, 1556 Copenhagen V
Denmark

info@mtp.dk | www.mtp.dk | tel. +45 3234 1414
VAT no.: DK 8876 8418

Guidelines for authors can be found at www.mtp.dk/MoG

Orders
Books can be purchased online at www.mtp.dk, via order@mtp.dk, through any of
MTP's worldwide distributors, or via online retailers and major booksellers.

More information at www.mtp.dk/MoG

About the series
Monographs on Greenland | *Meddelelser om Grønland* (ISSN 0025 6676) has published scientific
results from all fields of research on Greenland since 1878. The series numbers more
than 350 volumes comprising more than 1,250 titles. In 1979 *Monographs on Greenland* |
Meddelelser om Grønland was developed into a tripartite series consisting of

Bioscience (ISSN 0106 1054)
Man & Society (ISSN 0106 1062)
Geoscience (ISSN 0106 1046)

The titles in the series were renumbered in 1979 ending with volume 206 and continued
from volume no. 1 for each subseries. In 2008 the original *Monographs on Greenland* |
Meddelelser om Grønland numbering was continued in addition to the subseries numbering.

Further information about the series, including addresses of the scientific
editors of the subseries, can be found at

www.mtp.dk/MoG